Diatoms

This reference book covers the ecological impact and biotechnological applications of diatoms. Diatoms contribute to 40% of primary productivity in aquatic ecosystems. They have a unique biomonitoring ability and sensitivity toward different pollutants in wastewater, and are thereby a tool for ecotoxicological monitoring. The book discusses the need for newer diatom-based indices for wastewater quality analysis. It describes the nutraceutical value of the myriad compounds like omega 3 fatty acids-DHA, polysaccharides, and phenols in diatoms. It also elaborates on the challenges in sustainable diatom biorefinery. This book is meant for microbiologists, biotechnologists, and marine biologists.

Diatoms

Ecology and Biotechnological Applications

Edited by
Archana Tiwari
Ashok Pandey
Pau-Loke Show
Parameswaran Binod

CRC Press
Taylor & Francis Group
Boca Raton London New York

CRC Press is an imprint of the
Taylor & Francis Group, an **informa** business

Designed Cover image: © Shutterstock ID 2243721885, Diana Will

First edition published 2024
by CRC Press
6000 Broken Sound Parkway NW, Suite 300, Boca Raton, FL 33487-2742

and by CRC Press
4 Park Square, Milton Park, Abingdon, Oxon, OX14 4RN

CRC Press is an imprint of Taylor & Francis Group, LLC

ISBN: 978-1-032-33256-7 (hbk)
ISBN: 978-1-032-34430-0 (pbk)
ISBN: 978-1-003-32211-5 (ebk)

DOI: 10.1201/9781003322115

Typeset in Times
by MPS Limited, Dehradun

Contents

Preface

This book is a compilation about diatom biology. The broad-spectrum coverage of the book can be evidenced through the plethora of covered topics. Diatoms are the primary producers in aquatic ecosystems and contribute toward 40% primary productivity wherein magnificent photosynthetic machinery and diverse pigments play a vital role. The biomonitoring ability of diatoms and their sensitivity toward different pollutants present in wastewater is yet another unique feature associated with diatoms. The teratological deformities in diatoms in response to anthropogenic disturbances are a reliable toll for ecotoxicological monitoring. Envisaging newer diatom-based matrices for wastewater quality analysis can open new paradigms on a global platform. Current developments in diatom biology have been explored in this book to equip readers with diversified understanding. There are nine chapters in the book, covering a wide range of topics pertinent to the biology of diatoms.

The key features related to diatom taxonomy and new insights are discussed in Chapter 1.

Chapter 2 highlights Diatom High Cell Density Culturing: Challenges and Prospects. In diatom research, one major research gap is the suitability of diatoms for enhanced biomass production, and this chapter efficiently addresses the lacuna and explains current challenges and future prospects. From an ecological perspective, diatoms hold a significant role, and Chapter 3 – Ecological Impact of Diatoms: Mitigation to Indicators deals with the role of diatoms in an ecosystem. Diatom community dynamics due to sedimentation changes in the Savannah River estuary are mentioned in Chapter 4. In Chapter 5 – Carbon Dioxide Mitigation by Diatoms – appropriate mitigation strategies are envisaged.

In an aquatic system, the diatoms are ubiquitous and play a vital role, and this aspect is well described in Chapter 6 – Aquatic Role of Diatoms: From Primary Producers and Aquafeeds. In Chapter 7 – Climate Change and Diatoms, the impact of climate change on diatoms is discussed. Chapter 8 – Diatoms: Systematics to Nanobiotechnology is related to the role of diatoms in nanobiotechnology, and Chapter 9 – Cultivation of Diatoms in Photobioreactors elaborately explores the role of photobioreactors in cultivation.

We are thankful to the publishers and Dr. Madhurima Kahali, for their constant support.

Archana Tiwari
Ashok Pandey
Pau-Loke Show
Parameswaran Binod

Author/Editor Biographies

Dr. Archana Tiwari is a Professor at Amity Institute of Biotechnology, Amity University, Noida, India. She obtained Ph.D. degree from the University of Allahabad and has been a gold medalist and a distinction holder in Botany. Her key research interests include phycoprospecting diatoms for wastewater remediation and high-value products. Her research interests are on algal biotechnology and has published 70 papers and 10 books, with h index of 20 and about 1500 citations.

Professor Ashok Pandey is a Distinguished Scientist at CSIR-Indian Institute for Toxicology Research, Lucknow, India, and Honorary Executive Director at the Centre for Energy and Environmental Sustainability–India. His research and technological development interests are in industrial & environmental biotechnology and energy biosciences. Professor Pandey has 16 patents, 120 books, ~1000 papers and book chapters, with h index of 131 and > 74,000 citations.

Professor Pau-Loke Show is a Professor in Biochemical Engineering in the University of Nottingham, Malaysia. His research interests cover upstream to downstream bioprocessing, microalgae technology, and circular economy. Professor Show has published over 550 research papers and more than 200 contributions to international congresses, three books, and a few patents, with h index of 82 and about 29,000 citations.

Dr. Parameswaran Binod is a Principal Scientist in the Microbial Processes and Technology Division of CSIR-National Institute for Interdisciplinary Science and Technology, Trivandrum, India. He obtained his Ph.D. in biotechnology from the University of Kerala, Thiruvananthapuram, India. Dr Binod's research interests are in the areas of microbial & enzyme technology and bioprocess technology. He has published about 400 papers and book chapters with h index of 58 and about 13,000 citations.

Contributors

Mohd Azwan Ahmad
Department of Diagnostic and Allied Health Science
Faculty of Health and Life Sciences
Management and Science University
Selangor, Malaysia

Murni Amirra Mohd Aminuddin
Department of Environmental Health
Faculty of Health Sciences
MAHSA University
Jenjarom, Selangor, Malaysia

Jelena Z. Andrejić
Department of Biological and Environmental Sciences
Georgia College & State University
Milledgeville, USA
and
The University of the Bahamas
School of Chemistry, Environmental and Life Sciences, Oakes Field Campus
Nassau, The Bahamas

E.A. Bergey
Oklahoma Biological Survey and Department of Biology
University of Oklahoma
Norman, OK, USA

Rahul Bose
Phycology and Algal Biotechnology Laboratory
Department of Botany
University of Calcutta
35, Ballygunge Circular Road
Kolkata, West Bengal, India

K. Chandrasekhar
School of Civil and Environmental Engineering
Yonsei University
Seoul Republic of Korea

W. Cheah
Institute of Ocean and Earth Sciences
Universiti Malaya
Kuala Lumpur, Malaysia

Wai Yan Cheah
Centre of Environmental Health and Safety
Faculty of Health Sciences
UiTM Cawangan Selangor Kampus Puncak Alam, Puncak Alam
Selangor, Malaysia
and
Centre of Research in Development, Social and Environment (SEEDS)
Faculty of Social Sciences and Humanities
Universiti Kebangsaan Malaysia
UKM Bangi, Selangor Darul Ehsan
Malaysia

Kit Wayne Chew
School of Chemistry
Chemical Engineering and Biotechnology
Nanyang Technological University
62 Nanyang Drive
Singapore, Singapore

Wen Yi Chia
Department of Chemical and Environmental Engineering
Faculty of Science and Engineering
University of Nottingham Malaysia
Jalan Broga, Semenyih, Selangor Darul Ehsan
Malaysia

C. J. C. Derek
School of Chemical Engineering
Engineering Campus
Universiti Sains Malaysia Nibong Tebal
Penang, Malaysia

H. Fazhan
HICoE, Institute of Tropical Aquaculture and Fisheries
Universiti Malaysia Terengganu
Terengganu, Malaysia

Nur Azalina Suzianti Feisal
Department of Diagnostic and Allied Health Science
Faculty of Health and Life Sciences
Management and Science University
Selangor, Malaysia

Divyanshi Gupta
Department of Plant Science
Faculty of Applied Sciences
MJP Rohilkhand University
Bareilly

Mahima Gupta
Department of Plant Science
Faculty of Applied Sciences
MJP Rohilkhand University
Bareilly Bareilly

Taejun Han
Bio Environmental Science and Technology (BEST) Lab
Ghent University Global Campus
Songdomunhwa-ro, Incheon, Korea
and
Department of Animal Sciences and Aquatic Ecology
Ghent University
Ghent, Belgium

Tengku Nilam Baizura Ibrahim
Occupational Health and Safety Risk Management (OHSeRM) Research Initiative Group & Department of Environmental Health
Faculty of Health Sciences
Universiti Teknologi MARA
Cawangan Pulau Pinang, Kampus Bertam, Kepala Batas, Pulau Pinang
Malaysia

S.D. Ishak
HICoE, Institute of Tropical Aquaculture and Fisheries
Universiti Malaysia Terengganu
Terengganu, Malaysia

Noor Haziqah Kamaludin
Occupational Health and Safety Risk Management (OHSeRM) Research Initiative Group & Center of Environmental Health & Safety
Faculty of Health Sciences
Universiti Teknologi MARA
Puncak Alam, Selangor, Malaysia

N. A. Kasan
HICoE, Institute of Tropical Aquaculture and Fisheries
Universiti Malaysia Terengganu
Terengganu, Malaysia

Abhishek Kumar
Department of Plant Science
Faculty of Applied Sciences
MJP Rohilkhand University
Bareilly

S.S. Lam
HICoE, Institute of Tropical Aquaculture and Fisheries
Universiti Malaysia Terengganu
Terengganu, Malaysia

H. Manan
HICoE, Institute of Tropical Aquaculture and Fisheries
Universiti Malaysia Terengganu
Terengganu, Malaysia

Kalina M. Manoylov
Department of Biological and Environmental Sciences
Georgia College & State University
Milledgeville, USA

Thomas Kiran Marella
Algae Biomass and Energy Systems R&D Center (ABES)
University of Tsukuba
Tsukuba, Ibaraki, Japan

Sanjeev Kumar Maurya
Department of Plant Science
Faculty of Applied Sciences
MJP Rohilkhand University
Bareilly
and
Constituent Government College Puranpur Pilibhit

Sudeeksha Negi
Department of Plant Science
Faculty of Applied Sciences
MJP Rohilkhand University
Bareilly

Ruma Pal
Phycology and Algal Biotechnology Laboratory
Department of Botany
University of Calcutta
35, Ballygunge Circular Road
Kolkata, West Bengal, India

Lalit Kumar Pandey
Department of Plant Science
Faculty of Applied Sciences
MJP Rohilkhand University
Bareilly

Jihae Park
Bio Environmental Science and Technology (BEST) Lab
Ghent University Global Campus
Songdomunhwa-ro, Incheon, Korea
Center for Environmental and Energy Research
Ghent University Global Campus
Songdomunhwa-ro, Incheon, Korea
and
Department of Animal Sciences and Aquatic Ecology
Ghent University
Ghent, Belgium

Nur Faiza Abdul Razak
Department of Environmental Health Faculty of Health Sciences
MAHSA University
Jenjarom, Selangor, Malaysia
and
Technical Foundation Section
Institute of Medical Science Technology
Universiti Kuala Lumpur
Kajang, Selangor, Malaysia

S. Salleh
Center for Policy Research and International Studies
Universiti Sains Malaysia
Penang, Malaysia

Sangeeta
Department of Plant Science
Faculty of Applied Sciences
MJP Rohilkhand University
Bareilly

Pau-Loke Show
Department of Chemical Engineering, Khalifa University, Abu Dhabi, United Arab Emirates
Faculty of Science and Engineering
University of Nottingham Malaysi, Jalan Broga, Semenyih, Selangor Darul Ehsan
Malaysia

Meenakshi Singh
Department of Ecology & Biodiversity, Terracon Ecotech Pvt. Ltd.
Mumbai, Maharashtra, India

Pankaj Kumar Singh
Diatom Research Laboratory
Amity Institute of Biotechnology
Amity University
Noida, Uttar Pradesh, India

Shivangi Singh
Department of Plant Science
Faculty of Applied Sciences
MJP Rohilkhand University
Bareilly

Doris Ying Ying Tang
Department of Chemical and Environmental Engineering Faculty of Science and Engineering
University of Nottingham Malaysia
Jalan Broga, Semenyih, Selangor Darul Ehsan, Malaysia

C. Y. Tong
School of Chemical Engineering
Engineering Campus
Universiti Sains Malaysia Nibong Tebal
Penang, Malaysia

Rashi Tyagi
Diatom Research Laboratory
Amity Institute of Biotechnology
Amity University
Noida, Uttar Pradesh, India

K. Waiho
HICoE, Institute of Tropical Aquaculture and Fisheries
Universiti Malaysia Terengganu
Terengganu, Malaysia

Makoto M. Watanabe
Algae Biomass and Energy Systems R&D Center (ABES)
University of Tsukuba
Tsukuba, Ibaraki, Japan

1 Diatom Taxonomy

New Insights

Wen Yi Chia and Doris Ying Ying Tang
Department of Chemical and Environmental Engineering, Faculty of Science and Engineering, University of Nottingham Malaysia, Jalan Broga, Semenyih, Selangor Darul Ehsan, Malaysia

Kit Wayne Chew
School of Chemistry, Chemical Engineering and Biotechnology, Nanyang Technological University, 62 Nanyang Drive, Singapore, Singapore

Pau-Loke Show
Department of Chemical Engineering, Khalifa University, Abu Dhabi, United Arab Emirates, Faculty of Science and Engineering, University of Nottingham Malaysia, Jalan Broga, Semenyih, Selangor Darul Ehsan, Malaysia

CONTENTS

1.1 INTRODUCTION TO DIATOMS

Marine diatoms first appeared roughly one hundred million years ago and provide the majority of the organic stuff that sustains current sea life. The Greek word *diatomos* refers to the name of diatom, and it means "cut in half." It refers to their two-part silica cell walls. They live inside a diluted world where critical components for growth are shared and recycled, and the diatoms also have a significant impact on global weather, carbon dioxide concentrations in the atmosphere, and the function of marine ecosystems. Studies of these crucial creatures' genomes reveal how they will answer to the fast-changing conditions in oceans today, which is crucial for the environment's health (Armbrust, 2009). A diatom is a photosynthetic organism with a single cell, which indicates it will generate the foodstuff by itself in the identical manners as plants perform. At present, there are not less than 100,000 species of diatoms in total from above 1,250 genera with the diversity surface structure of diatoms in this world (Kooistra et al., 2007; Zhang et al., 2019). Diatoms are a large group of algae that construct one of the most ordinary phytoplankton types. Out of many ocean organisms, they float on the ocean and lakes' upper layers on currents and can be discovered

DOI: 10.1201/9781003322115-1

everywhere, such as oceans, lakes, rivers, moist rock surfaces, bogs, and the surface of the skin on a whale (Seckbach & Kociolek, 2011). Besides that, diatoms are a diverse, abundant, and significant algal assemblages' part in lakes with freshwater. They make up a significant amount of total biomass of algae across the lake's trophic spectrum (Hall & Smol, 2010).

Other than that, diatoms are also classified as Class Bacillariophyceae and are a principal group of eukaryotic, siliceous algae that take part in an important role in fixing carbon dioxide, synthesizing carbohydrates, and maintaining ecosystem health. They contribute around 20%–25% of global primary production even though their individual size is less than 200 μm (Rühland et al., 2015; Saade & Bowler, 2009). Diatoms have a vital role in the structure and operation of the aquatic food web since they are a plentiful, high-grade food source for herbivores (Christensen, 1991). Diatoms are commonly employed in palaeoecological investigations due to the huge ecologically sensitive species amount that conveniently leave a fossil record (Bayer & Mann, 2000).

Most aquatic ecosystems' primary and secondary consumers rely on diatoms to some extent and even completely (Lundholm et al., 2018). Many fish, shrimp, and shellfish species with high commercial value feed primarily on diatoms (Kumaran et al., 2016). As a result, diatoms represent the primary carbon source for a diverse spectrum of aquatic food webs, as well as a significant share of global fisheries (Jackson et al., 2010). Therefore, diatoms also support the marine food web and aid in carbon export from the surface to the depths of the ocean. They act as part of the biological pump and account for around 40% of primary production of marine and particulate carbon exported to depth. Furthermore, the diatoms are plentiful in nutrient-rich waters, but they are also prominent in meso- and sub mesoscale features like fronts and filaments as well as in the deep chlorophyll maxima based on observations and simulations (Tréguer et al., 2018). Diatoms' wide-reaching ecological success has proposed that they have honed their cellular processes to a higher level than most of the other unicellular algae for effective utilization of sunlight and nutrients. Moreover, the size, chemical composition, and morphology of diatoms will influence the quantity, quality, and biogenic materials' sinking speed to depth (Sharma et al., 2021). Particularly, their silica shells give balance to faecal pellet and marine snow and can aid in the transport of carbon to both the upper and mesopelagic layers of the ocean (Hamm et al., 2003).

Furthermore, diatoms are distinctive due to their silicified cell walls that give mechanical protection from grazers among the microscale planktonic primary producers. Diatoms have also become the most plentiful and various siliceous marine microorganisms, and they have come out even as the silicon cycle main factor (Finkel et al., 2005; Tréguer & Rocha, 2013). The large size and fast growth rate of diatoms in an enriched nutrient environment make them a crucial contributor to organic carbon production in the euphotic zone (Pierella Karlusich et al., 2021). The organic carbon produced by photosynthesis on the ocean surface is transported to the trophic network and subsequently exported to depth, where the majority of it is remineralized to form carbon dioxide. The biological carbon pump is a phrase for this type of export (Tréguer et al., 2018).

Finally, diatoms have a broad range of possible applications in environmental research and management such as being used as environmental indicators because of their high sensitivity and preservability. For instance, the diatoms will modify their individual morphology or size and assemblage according to changes in the environment (Hinder et al., 2012; Li et al., 2018). Next, diatoms can be used to recreate paleoenvironmental history and even population dynamics of specific aquatic species (Li et al., 2018). Not only that, but the diatom silica shells are also easily preserved and make excellent fossil study tools (Wang et al., 2014). Promoting research in a particular field necessitates an understandable and objective picture of its evolution history and current state. Nevertheless, a comprehensive summary of research in diatoms is still lacking (Zhang et al., 2019).

1.2 DIATOM TAXONOMY AND IDENTIFICATION

For diatom-based water quality research and biotechnological applications, diatom taxa must be correctly identified. The identification of species from specific specimens is one of the biological applications of taxonomy. Due to the enormous biological variety seen in this category of microalgae, identification keys, which utilize a preset set of traits to recognize a specific individual, are rarely used by diatom specialists.

The development of modern identification tools, such as numerical taxonomy, electron microscopy, metabarcoding, and so on, make taxonomy an interesting study topic in the biological sciences (Leuschner & Sviridov, 1986). Traditional taxonomy relied on intuitive or subjective similarity judgments, which were largely dependant on experience and taught abilities (Dunn & Everitt, 2004). Despite the fact that identification keys have long played an important role in biological identification methods, their use in diatomology is less common than in other fields of botany (Clark & Warwick, 1998). Identification of microalgae is typically based on the specimen's "overall appearance" or certain important features, as well as visual comparison with reference floras and matching specimens with images (Walter & Winterton, 2007). When the number of taxa increases at an exponential rate, for instance in the case of diatoms, this strategy rapidly becomes intractable. Given that a typical freshwater sample may include tens of distinct species, the "traditional" identification process necessitates a significant amount of labor to identify diagnostic features, which clearly has a significant role for diatomists (Zhang et al., 2008).

Biologists built keys to plants and animals more than two centuries ago, and the mechanisms for doing so were debated far earlier (Reynolds et al., 2003). Keys are essentially a series of questions ("tests") regarding the characteristics displayed by the organism for identification, with alternatives based on the outcomes of previous replies. In contrast to visual matching, the usage of keys guarantees that all important characters are analyzed, making the identification process much more objective (Reynolds et al., 2003; Talent et al., 2014). Although strategies for the automatic production of identifying keys are published in the literature, diatomists often construct their keys manually (Talent et al., 2014). The use of computers results in ideal identification keys that reduce the amount of tests required, but only a few tools can assist in the generation of identification keys without the supervision of an expert (Payne & Preece, 1980; X.-b. Zhang et al., 2008).

The technique usually starts with the generation of a data matrix (X taxa Y characters), with entries in each column indicating the value that each character takes for each taxon (Blanco, 2020). A vital step is selecting appropriate diagnostic characteristics. The optimal trait should combine simplicity of observation with high information content (the ability to separate taxa, see (Blanco, 2020)), that is, characteristics that have the highest consistency within taxa yet divide the collection of taxa under investigation into about equal halves (Dunn & Everitt, 2004). Characters with both characteristics should be inserted in the initial stages of the key (Calvo-Flores et al., 2006). In the case of diatoms, qualitative traits (especially valve outline) are difficult to describe and more prone to confuse, whereas quantitative characters (valve size, stria density) show constant change, making taxonomy identification challenging (Talent et al., 2014). Other important diagnostic character features are (1) life-cycle persistence, (2) existing in all individuals, and (3) independence from the status of other characters (Kale & Karthick, 2015). Diatom identification is frequently hampered by the presence of detritus, incomplete forms of individuals, or an extremely small community that prevents observation of size ranges or various viewpoints (valve/connective views). The taxonomic significance of colony type is usually lost during slide processing.

Taxonomic identification poses unique challenges in diatomology. Observing a full population is frequently required to account for the entire morphologic or morphometric diversity of the taxon with taxonomic importance; instead, isolated individuals (valves or frustules) cannot be used to make species-level conclusions (see in this chapter). Some keys deal with the existence of quantitative characters by assigning a single representative value (e.g., the mean or the median) to each changing character, resulting in data loss (Karbalaie et al., 2021). Furthermore, for some species, the presence of individuals from diverse perspectives (valve/connective view) may be essential. The occurrence of "species complexes" (Tapolczai et al., 2021) or "cryptic taxa," which are challenging to distinguish even for specialists, further complicates identification in this group. When such crucial tests provide ambiguous results, complete identification cannot be ensured since numerous species might show the same result (Hu et al., 2021). If precise identification is unattainable, either because too many characteristics vary between taxa or because all evaluations of character states are vulnerable to mistake, probabilistic identification approaches can be utilized (Calvo-Flores et al., 2006). Based on how frequent specific character states occur in taxa and the taxonomy's own relative rates of occurrence in some cases, these keys can create a probability value for each determination (Payne & Preece, 1980).

Another issue is that taxonomic keys for algae are frequently confined, and it is never clear to which extent such keys can be transferred to larger or more distant geographic areas (Tapolczai et al., 2021). For the reasons described above, taxonomic keys are rather sparse in current diatom literature. They are absent from the majority of books in the Bibliotheca diatomologica and Iconographia diatomologica collections. Diatoms of Europe, which offers a detailed description of known species classified by genera from all over the world, provides significant "visual keys" for the initial delimitation of morphological groupings.

Accurate, high-quality images are essential in any taxonomic investigation. Phycologists have employed pictures to define nearly all reported species since diatoms have been acknowledged as a unique class of aquatic microorganisms. Illustrations are useful not only for morphological or aesthetic purposes, but also for accounting for the differences that allow species classification and providing evidence for the morphologic/morphometric variety of the groups studied. Illustrations are defined in botany as any "piece of art or picture portraying a feature or characteristics of an organism" (Turland et al., 2018), which includes drawings and microscopic images ("micrographs"). New diatom taxa must go along with an illustration exhibiting their unique structural characteristics in order to be legitimately published. Illustrations were also used to typify names, which is to give element(s) to which the species name is permanently connected. An automatic annotation approach was developed to perform the image segmentation of diatoms (Figure 1.1) (Dimitrovski et al., 2012). The procedure starts by separating the touching objects, then creating the basis for where the light pixels are high and dark pixels are low. This watershed segmentation method can isolate the pixels that should be removed and extract the main diatom contours.

1.3 DIATOM BIOLOGY AND CLASSIFICATION

Apart from taxonomy, the life cycle of diatoms is also one of the interesting topics discussed by researchers. This is owing to the alternation of meiotic and mitotic stages, which last for months and days (Mann, 2011). The valve shape of the diatom also changes significantly when the smaller cells are generated (Figure 1.2) (Loke et al., 2004). Thus, the auxospore, vegetative, and sexual phases are comprised in the life cycle of a diatom. At the meiotic stage (sexual phase), the centric diatoms are differentiated into several male and female cells, which are spermatocytes and oogonia, respectively. This is then followed by the formation of auxospore via sexual production where the spermatocytes fertilize the female oogonia through the opening in the oogonia thecae (Moore et al., 2017). The auxospore formed then expands to maximum and breaks open to form an initial cell and cell valve (Bidgood et al., 1999). This initial cell then enters the vegetative phase, where vegetative growth and cell size reduction will be performed. As such, mitosis is carried out in which the mean cell size is significantly reduced. The diatom cell will divide by cytokinesis, and the new valves will form. On top of the intact wall layer of the parent cell, the dividing cell remains highly rigid, which resulted from the silica impregnation. Therefore, one of the daughter cells will be in the size of the parent cell while others will be smaller (Joseph & Patrick, 2011). In the meantime, the cell cannot be sexually induced in the refractory period until it exceeds the sexual size threshold under favorable environmental conditions. Then, the cell will proceed to the meiosis stage, where the cycle will be repeated.

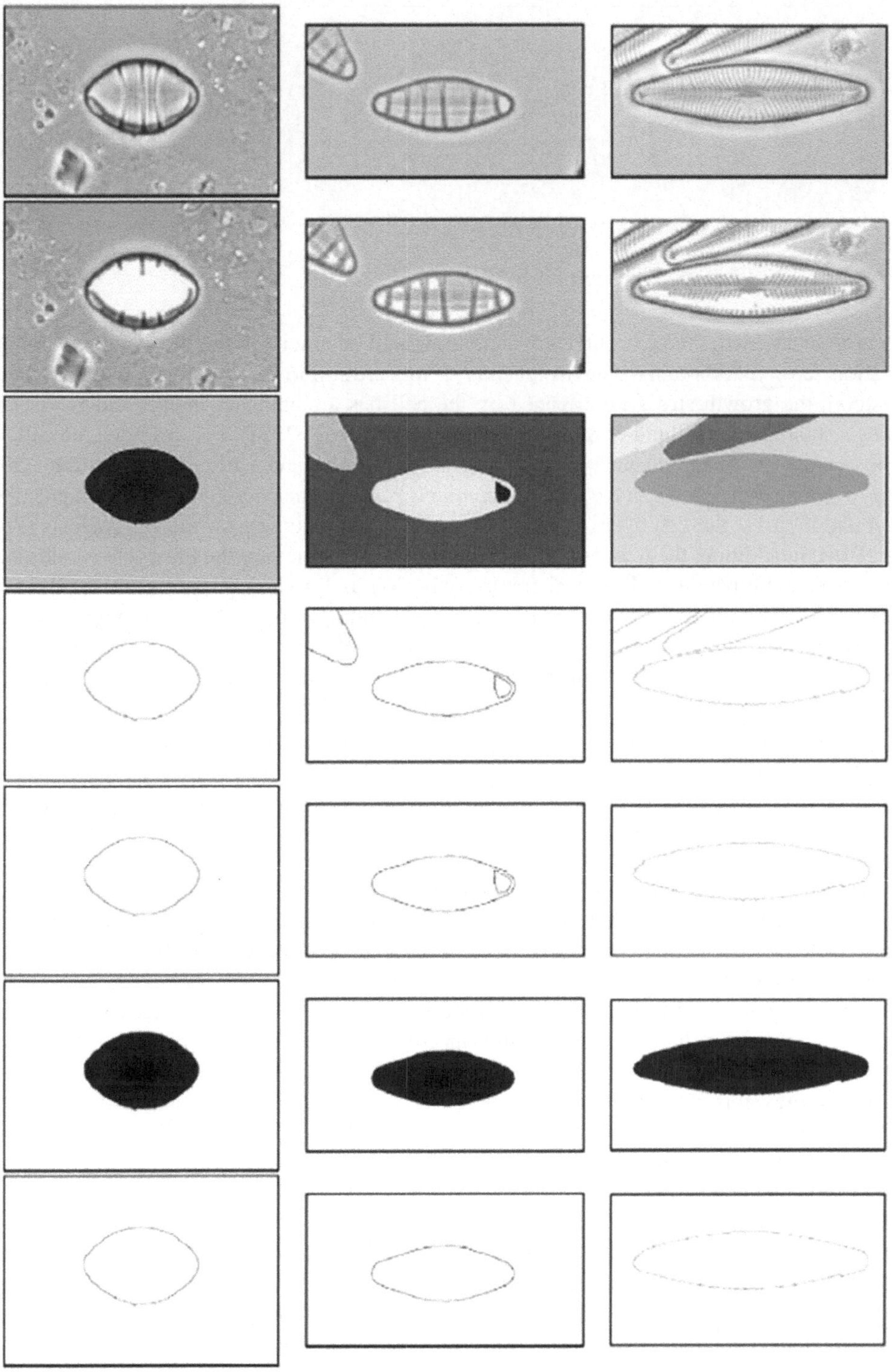

FIGURE 1.1 Schematic diagram of a diatom segmentation procedure for three diatom species (from left: *Diatoma mesodon*, *Denticula tenuis,* and *Navicula lanceolate*). The process steps from top to bottom: Imaging with markers, imaging with watershed regions, imaging with watershed lines, imaging with contours, regions after flood-filling the contours, and imaging of the final diatom contours (Dimitrovski et al., 2012).

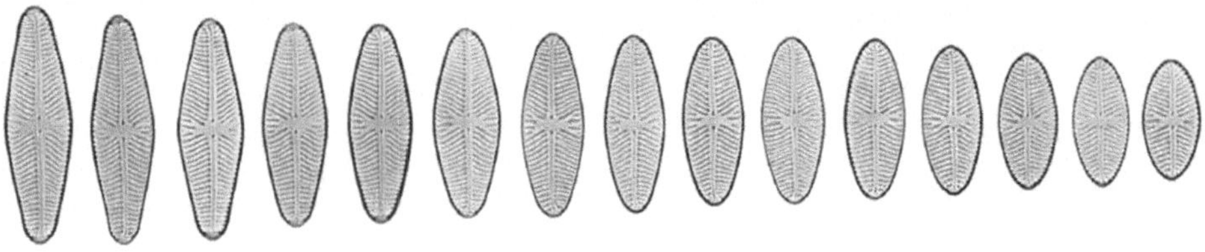

FIGURE 1.2 The life cycle of the diatom (Loke et al., 2004).

As aforementioned, the environmental conditions will be crucial in the life cycle, including the population density. Although the environmental factors are still in the investigation phase, yet to be understood, the growth stress experienced by the cell has a significant impact and is associated with the sexualization (Edlund & Stoermer, 1997; Moore et al., 2017). For example, the cell cycle will be stagnant with the changes in light intensity, salinity, and nutrient composition (Godhe, Kremp, & Montresor, 2014). In the case of *Seminavis robusta*, the pheromons from pennate diatom were found to arrest the cell cycle, thus inducing the sexual pathway for the cell meiosis (Moeys et al., 2016). In addition, the temperature and duration also might vary the life-cycle condition. For instance, the combination of low temperature (<14 °C) and a short period promotes the sexual phase in the *Cocconeis scutellum,* which is a marine benthic diatom, whereas a longer period at a high temperature ranging from 14 °C to 18 °C is preferable for vegetative growth of *Cocconeis scutellum* (Mann, 2011). On the contrary, the nutrient, N, content in the culture medium will inhibit the auxosporulation of *Pinnularia nodosal* (Poulíčková & Mann, 2008).

Although it is always assumed that the vegetative phase will be started after auxosporulation, followed by the sexual phase (meiosis and fertilization), there is no absolute relationship between the sexual phase and auxosporulation. This is because the unnoticeable fusion of spermatocytes in oogonia causes the suspect of the asexually produced auxospore in the case of a centric diatom. Additionally, the araphid pennate diatom is also considered to have asexually produced auxospores in the study of Sato and his group (Sato et al., 2008). Meanwhile, the clear and obvious link between fertilization and auxosporulation can be found in the raphid diatom as the visible copulation between the vegetative cells can be obtained in the diatom (Joseph & Patrick, 2011).

Generally, diatoms are formed by the frustule, which consists of the epitheca and hypotheca. On top of the shape of the frustule, diatoms can be categorized into centric diatoms (centrales) and pennate diatoms (pennales) (Moore et al., 2017). The frustule in the former is rounded with radical markings and immobile, whereas the latter is elongated with pinnate markings, which mostly can be found in freshwater (Sánchez et al., 2019a; D. M. Williams & Kociolek, 2011). The former can then be further divided into a few sub-categories, such as *Coscinodiscineae*, *Rhizosoleniineae*, and *Biddulphioideae* (Joseph & Patrick, 2011). For the latter, the features of the pennate diatom are represented in Figure 1.3 (Salido et al., 2020). The *Sellaphora pupula* complex composes six variants that are very difficult to separate using morphology itself. Hence, the use of automated feature extraction and identification algorithms would be required to aid in the taxonomic and ecological study of this complex diatom. The six variants (demes) of the *Sellaphora pupula*, representing the fine-grain morphological variation, are shown in Figure 1.4 (Loke et al., 2004).

On top of the distinct types of diatoms, the classification technique becomes important in identification steps. In the past, morphometric parameters like length, width, and frustule density would be used in the identification by comparing with the known specimen. However, it becomes challenging when the number of diatom species gets larger. Also, the classification might be affected by the changes in the shape and texture during the life cycle. Herein, there are two

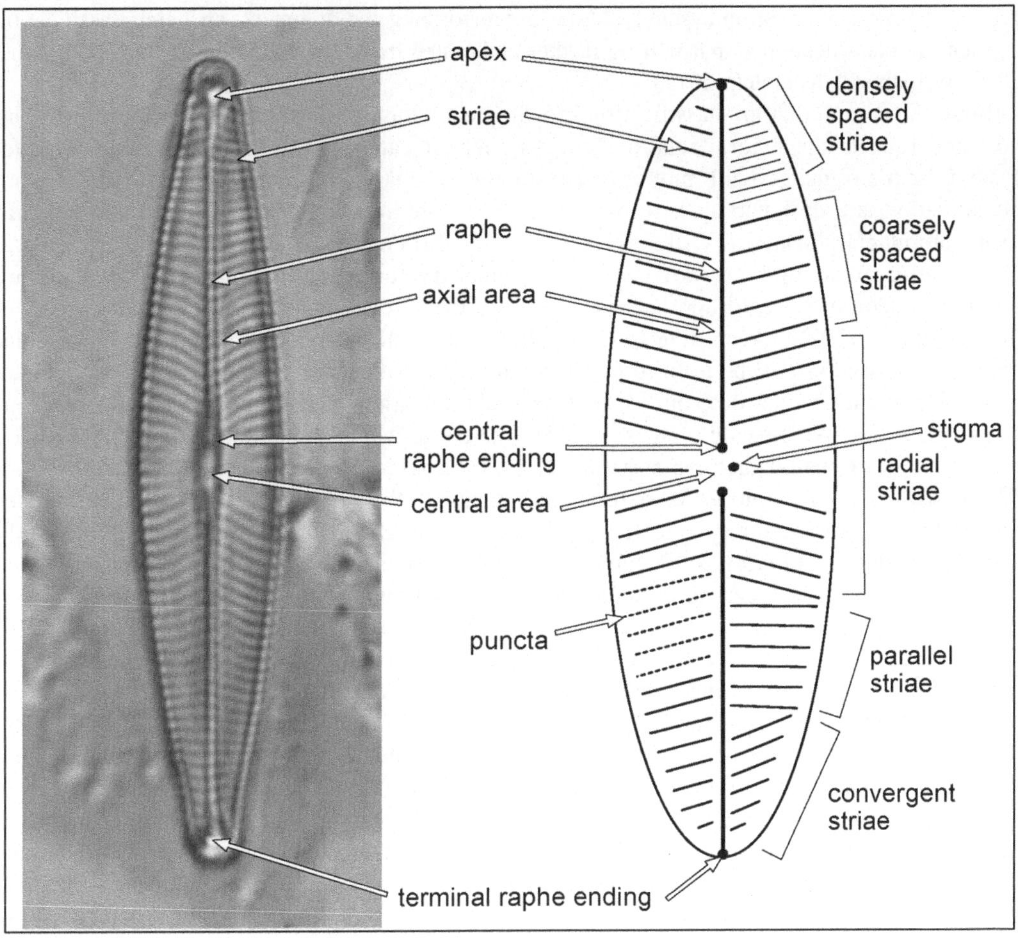

FIGURE 1.3 Features of a pennate diatom valve (Salido et al., 2020).

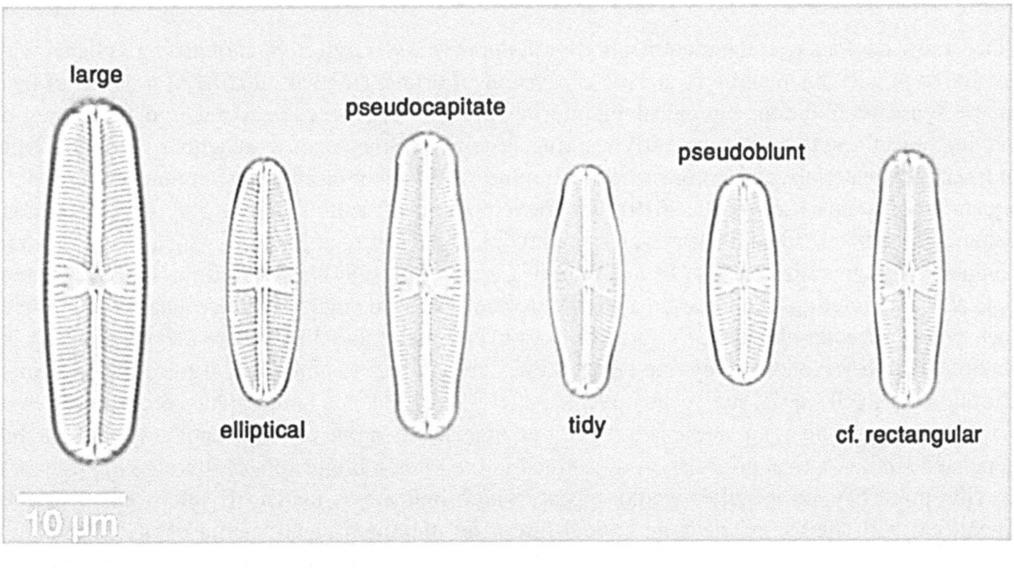

FIGURE 1.4 Classifications of six demes of *Sellaphora* pupula and its provisional names (Loke et al., 2004).

general diatom classification systems aiding in categorizing the diatoms – artificial classification and natural classification – that will be further elaborated on in the following section (Williams, 2020; Williams & Kociolek, 2011).

First, the natural diatom classification is a simple attempt to illustrate the relationship among taxa and reflects the natural relationship. As such, it can serve as an information retrieval system for the organism with phylogenetic studies as a base. Meanwhile, the artificial diatom classification is a designed system to facilitate recognition, thus making it much more useful than natural classification (Williams & Kociolek, 2011). Yet, it could not replace natural classification owing to the working principle. Under the former natural classification, there are two non-randomly selected approaches, including Cox's classification and Mann's classification (Williams, 2020). According to Cox (2015), the wall morphology and recent molecular phylogenetic studies will be utilized as a base in the classification. However, the low accuracy of the DNA cladograms determined by Cox's classification make it so the natural relationships may not be revealed (Cox, 2015), whereas Mann's classification will provide some basic information that can serve as the starting reference for the taxonomic framework (Williams, 2020). In contrast to natural classification, the artificial classification comprises randomly and non-randomly selected guidelines. Through the implementation of randomly selected guidelines (Taylor & Cocquyt, Diatomeas del Canal, and Atlas des diatomées), a simple comparison between specimens and images is provided, which will be useful in biomonitoring and outlining the species found in the area of interest (Roy & Karthick, 2019; Williams, 2020). In the meantime, there are three non-randomly selected guidelines found in the artificial classification, including Diatoms of the United States (DUS), Diatom Flora of Britain and Ireland (DFBI), and Freshwater Benthic Diatoms of Central Europe (FBDE). The former, DUS, can aid in complying with the Clean Water Act as detailed information like taxonomy, ecology, and nomenclature will be provided for the diatoms, whereas the latter, DFBI and FBDE listed the genus descriptions alphabetically to provide the number of inclusive genera for referring purposes (Williams, 2020). In short, there are two general classifications for diatoms; the natural classification reveals the taxonomy and relationships of diatoms, whereas the artificial classification provides detailed information to boost the recognition of diatoms.

1.4 TRENDS IN DIATOM IDENTIFICATION

Since a few decades ago, the identification of diatoms is associated in biomonitoring systems with the aim to protect the integrity of aquatic ecosystems (Forio & Goethals, 2020). As a matter of fact, in the sense of tradition, biological monitoring systems can be categorized into two types of varying objectives: to serve as an early warning system to instantaneously diagnose industrial spills of hazardous materials; and to consistently examine the health of biological communities present in aquatic ecosystems (Lei et al., 2019). The development of a biomonitoring system for diatom community structure identification was essential due to the nature of diatoms, which might result in complex shifts in water quality. In this regard, Cairns et al. (1977) proposed the use of coherent optical spatial filtering techniques in a biomonitoring system to quickly identify diatoms along with their process species-abundance information. Conventionally, the diatoms are identified primarily according to the taxonomic key that demonstrates certain patterns of physical form in the cleared frustule of the cell. In the study, the optical spatial filter acted as a holographic record of the two-dimensional periodic or aperiodic structure of an object. When one of the unknown specimens that generated similar optical information was stored in the filter, a bright spot of illuminating light at a specific point beyond the filter would appear, which indicates a match of pattern has occurred. Therefore, with the use of different optical filters, an unlimited identification of diatom patterns could be made possible as any species of diatom has an exclusive set of structural properties that are favorable in this case. In the end, the author claimed that the diatoms with desirable properties of rigidity of form and complex character sets could be suitably determined by using coherent optical spatial filtering techniques (Cairns et al., 1977).

Over the years, with the advent of microscopes, the studies related to diatoms have progressed rapidly, and more researchers have switched their interest of study in investigating the details of morphological features of diatoms. Owing to the advancement of diatom research, various metrics associated with diatom cell wall such as length, breath, striae density, and the density of wide array of valve features were included in the study to describe the large number of diatom species (Pappaset al., 2014). In fact, the biometric or traditional morphometric methods were implemented with the aim to study the diatoms' variation. Pappas et al. (2014) reviewed the history of using morphometric methods along with the outline of shape analysis methods as well as the application of new geometric morphometric methods in diatom research. The authors concluded that with the introduction of computing power into these methods, an entirely separated study that is related to 3D applications can be accomplished to further explore the morphology of diatoms for taxonomy and identification investigations (Pappas et al., 2014).

The traditional methods for identification of diatoms are, indeed, significantly relying on the taxonomy of diatoms, which might face three problems:

1. The diversity of diatoms is unknown but undeniably more enormous than ever expected (Mann & Droop, 1996).
2. The identification of diatoms is getting more and more challenging as the existing taxonomic aided technologies are rendered obsolete whenever new discoveries are found (Evans & Mann, 2009).
3. Contrary to the long-held belief that all taxa are cosmopolitan, species, and even genera, can possess biogeographies (W. H. Kooistra et al., 2008).

The dependence of diatom taxa characterization and identification on valve morphology was claimed to be an inadequate basis for future work due to the vast cryptic diversity (Zimmermann et al., 2015). As a result, these issues have become the prime candidates for the development of DNA barcoding (for single specimens, metabarcoding for multiple specimens) and eDNA metabarcoding (for environmental samples) in the identification of diatoms from the molecular data (Blanco, 2020). In this regard, Evans and Mann (2009) presented the mechanism of DNA barcode nomenclatural status in microalgae via culture-derived epitypes. This study illustrated the use of barcodes in stabilizing taxonomy by giving them nomenclatural status via holotypes or epitypes (Evans & Mann, 2009). Other than that, Hamsher et al. (2011) studied the alternatives to COI-5P, a common region used for DNA barcoding, in terms of the universality. Different markers have been applied including ~1400 bp of *rbc*L; 748 bp at the 3' end of *rbc*L (*rbc*L-3P); LSU D2/D3 and UPA to examine each marker's ability to discriminate among closely related species of *Sellaphora* isolates. The results showed all the alternative markers demonstrated greater universality than COI-5P and suggested that *rbc*L-3P was suitable to be used as the primary marker for diatom barcoding, whereas LSU D2/D3 should be used as a secondary marker in assisting environmental surveys (Hamsher et al., 2011). To sum up, the use of the "taxonomy-free molecular diatom index" in diatom identification is crucial to get rid of the problems related to the diverse lineage of diatoms (Blanco, 2020).

As aforementioned, diatoms are commonly utilized in monitoring the water quality. However, the traditional identification methods are difficult, are time-consuming, and require complicated knowledge, as the non-homoplastic species-specific morphological characters are scarce. Hence, Zimmermann et al. (2015) worked on the comparison between the identification methods using DNA (metabarcoding using next generation sequencing, NGS platforms) and morphological techniques. The authors also proposed a workflow in optimizing the diatom fresh water quality assessments. In this study, seven different sites of the river that contributed to diatoms diversity were analyzed by using DNA metabarcoding and morphological methods. The findings indicated that NGS technology was capable of identifying 270 taxa as compared to light microscopy with 103 taxa. It was found that most of the rare taxa from oligotrophic parts of the river systems were unable to be identified due to the lack of information provided in the reference database (Zimmermann et al., 2015). In short, this study provides evidence that DNA metabarcoding using NGS sequencing for diatom identification is undoubtedly able to replace the old morphological methods for water quality assessments. Various investigations have also been accomplished to compare the diatom identification methods for environmental assessments between morphological and DNA metabarcoding (Apothéloz-Perret-Gentil et al., 2021; Mora et al., 2019; Mortágua et al., 2019; Rivera et al., 2018).

Recently, Zhou et al. (2020) carried out a review in identifying and testing the presence of diatoms in the water that originated from dead bodies. The main focus of the study was to investigate the possible sites of the drowning cases by comparing the diatom species abundance patterns from tissues with those collected from the suspect water samples. The authors also highlighted that DNA barcoding that utilized specific genetic makers in identifying particular species has been widely investigated in forensic applications. In a similar case, Zhao et al. (2018) proposed the use of pyrosequencing (PSQ) to execute multidimensional analysis of diatom signals in water samples. The correlation between the water samples and collection locations could be obtained via multidimensional analysis of the PSQ curve of the diatoms in the water samples (Zhao et al., 2018). However, the exorbitant cost of the DNA barcoding analysis has hindered its application in forensic practice, which has signified the importance in implementing other methods such as modified digestion methods that may increase the detectable rate of diatoms and maintain low practicing cost (Zhou et al., 2020).

In summary, the development of better, faster, more accurate methods for the identification of diatoms are constantly desirable (Blanco, 2020). Throughout the review of the trend of diatom identifications, it was found that most of the current studies are closely related to the water quality or aquatic environment pollution assessment due to the characteristic of diatoms that serve many ecological services (Valentin et al., 2019). Some of the studies are indeed starting to involve deep learning or machine learning in diatom identification so as to facilitate the automation system and subsequently enhance the performance of the methods (Feio et al., 2020).

1.5 NATURAL AND ENVIRONMENTAL DIATOM MORPHOLOGY

Diatoms can be classified based on many different criteria; morphology is one of them. Different morphology is related to the particular functions that are useful in their identification and classification (Fu et al., 2022). The never-ending evolution of diatoms has resulted in variations of shape/size observed in the diatom population. The diversity of diatoms resulted from either environmental factors or their particular asexual reproduction cycle (Adriana Olenici et al., 2020). Environmental factors play a significant role in inducing the stress for the diatoms to adapt and evolve. Diatoms are capable of adapting to the surrounding environment in both short-term and long-term manners. By altering their physiological and morphological traits in response to environmental stressors such as pH, temperature, salinity, light, resource availability, and grazing, diatoms have the flexibility to adapt to the surroundings (Fu et al., 2022). That adaptability is crucial for diatoms to dominate the phytoplankton population in the modern ocean. Mancuso et al. (Mancuso et al., 2021) studied the response of phytoplankton in Muskegon Lake in response to the unusually cool temperature. The results of biweekly sampling have suggested that diatoms were the single dominant division throughout the entire growth season. Out of 13 samples, 10 of them were dominated by the diatom populations by over 75% of the phytoplankton community. This is also an indication that the population of diatoms bloom in spring survived through the fall.

Attention has been invested on a teratology study revolving around diatoms as the development of teratological forms within the microalgal population emerges. Teratological forms are the non-adaptive phenotypic anomalies generated by environmental stresses (Falasco et al., 2009). Teratology has been detected in the laboratory and infield and can be induced in a diatoms culture (A Olenici et al., 2019).

Diatom teratological forms are one of the most effective individual-level indicators because they respond rapidly (approximately 1 week) to a variety of environmental stresses, including novel-developing contaminants (Falasco et al., 2021). Response toward the contaminant generally takes a week, which is fast enough to recover the ecosystem health status, preventing disturbance toward the following organization level (e.g., population, community, and food chain). On top of being an early detection system, diatoms are also used to track past environmental conditions.

Valve deformations are primarily influenced by trace metals and acid conditions (da Silva et al., 2009; Leguay et al., 2016), but others have suggested other variables, such as high content of ammonium nitrogen and phosphorous (Koreneva & Latkovskaya, 2013; Majewska et al., 2012), artificial growth conditions (Falasco et al., 2009), mechanical stress, or parasitism (Munnecke et al., 2012). Metal contaminations on the teratological valve have been widely discussed, but the root cause of biochemical mechanisms is poorly known (Adriana Olenici et al., 2020). Metals disrupt the biochemical and cellular processes, resulting in cell malfunction that limits the growth and may eventually lead to cell death (Masmoudi et al., 2013). On top of altering the valve features, pollutants are also responsible for nuclear abnormalities, changes in photosynthetic apparatus, and lipid bodies (Pandey et al., 2017). Contrastingly, Falasco et al. (2021) suggested that both cytoskeleton and silicon metabolic pathway abnormalities are likely implicated in teratological valve mechanisms and that the teratology do not appear to deteriorate the reproduction capacity and viability of the affected individuals. Although it was hypothesized that cells affected by abnormalities are at a disadvantage and will eventually die, the viability of aberrant diatoms may be linked to the type of teratology. For example, the raphe disruption might reduce the viability and reproductive capability, particularly in natural environments where cell adhesion is critical (Falasco et al., 2009). On the other hand, striae and mixed deformations are likely to help cells survive and reproduce (Arini et al., 2013).

It was hypothesized that higher metal contaminations leads to more evident damage, resulting to lower viability. Lowered motility rate in severely Cu and Zn polluted sites were observed by Pandey and Bergey (2016). The diatoms in a less polluted area illustrated a straight curved or sigmoid path as opposed to the random and erratic paths in highly polluted area. However, Coquillé and Morin (2019) discovered a similar growth rate of *G. gracile* between teratological strains and normal individuals with a relatively comparable photosynthetic efficiency and mobility. This result was contradictory to the past experiment done by Falasco et al. (2021) using *G. gracile* strain with the same artificial condition. The recent discoveries have highlighted the evolutionary success of teratological forms, particularly for culture conditions in the natural habitat (e.g., rivers and streams). The appearance of teratological individuals is lower in natural conditions as compared to artificial conditions. In natural conditions, the decontamination process continuously takes place, whereas the presence of teratological individuals is covered by the continuous flow of normal diatoms coming from the upstream regions, functioning as a source (Coquillé & Morin, 2019). Migration of diatoms has been accepted as the main mechanism to recover the diatom community (Falasco et al., 2018).

At the intraspecific level (Luís et al., 2011) or condition with a dominance of small-sized diatoms in contaminated streams (Barral-Fraga et al., 2016), cell size reduction is commonly observed. The occurrence of size diminution is a common consequence of asexual reproduction cycle. At each cell division, diatoms will be reduced by half of frustule thickness, resulting in a continuing decline in average cell size to the point where auxospores can be generated, allowing maximal size to be restored (Hendey, 1951). While new valves conform to the pattern of the parent cell, once the parent frustule is discarded, the cell is free to rehabilitate with a frustule whose outline may not be similar to that of the pre-auxospore generation, as long as the essential design is preserved.

Valve shape remains stable across wide chronological scale; however, after several generations there is a decline in the valve size. Smaller valves are associated with better vegetative reproduction (Shukla & Romero, 2018). Variations of cells are possible (Meyer et al., 2001) as the external influences may modify the size of post-auxospore frustules (Lund, 1946), after auxospore formation has begun. Furthermore, due to the apical axis shortening quicker than the transapical axis, size reduction in diatoms is frequently accompanied by an allometric shift in valve proportions (Kociolek & Williams, 1987; Schmid, 1994). As a consequence, valve outline varies with time, with shorter valves looking broader than longer ones and various types becoming more rounded (Cox, 2010). Hence, most shape descriptors are substantially linked with valve length when determined for morphologically similar diatom taxa (Potapova & Hamilton, 2007).

Effects of diatom change in size are reflected in the physiological rates and ecological functions (Finkel et al., 2010). Diatom size mediates the intraspecific and interspecific interactions in the environment (Pfister & Stevens, 2002). Larger cells have the upper hand in gathering resources and adapting to predation (Giuseppe Cortese & Rainer Gersonde, 2007), and they are more heavily silicified as compared to smaller cells (Pappas & Stoermer, 2003). Additionally, diatom size limits the rate of obtaining and processing energy (Finkel et al., 2010), leading to the specious amount of smaller diatom sizes in plankton and bigger diatom sizes in benthos (Passy, 2007).

1.6 RECENT DIATOM DISCOVERIES

To date, there are more than 200 genera of diatoms known to science, with approximately 100,000 species (Kulkarni, 2020). Diatom species discovery today, as in the past, begins and frequently concludes with a survey of morphological variation to uncover deviations in the structure that allow distinguishable entities to be identified and named. Back to the beginning of the 18th century, an unknown English microscopist discovered and described diatoms with a drawing demonstrating *Tabellaria*, which is chains of cells that can be specifically identified. He said that each chain was made up of rectangular oblongs and perfect squares that were joined together where those rectangular parallelograms were all having the similar size and texture. These statements document traits of diatoms that have been significant for taxonomy since then, such as cell size, shape, and surface pattern as well as the retention of many morphological characteristics of the living organism by dead cell walls. Figure 1.5 shows the scanning electron micrographs (SEM) of a typical diatom frustule and the magnification of the silica biopatterning. It is vital to explore the mechanisms that enable diatoms to fabricate such a structure at ambient conditions, to further process it for applications in chemistry, nanotechnology, and materials science (Lopez et al., 2005). In fact, some taxonomically significant life-cycle aspects were not accurately documented until 1932 as many mistakes were made in the early 19th century due to a lack of understanding of the diatom life cycle and morphological diversity (Mann, 2010).

Diatom research extends beyond biodiversity discovery to a rich array of topics, including water quality and bioassessment, the effects of eutrophication and dam construction, paleoenvironmental reconstructions, and the development of a variety of diatom-based products (Kociolek et al., 2020; Ouyang et al., 2016; Shen et al., 2018; Zhang, 2019). In the context of biodiversity discovery, biogeography, and resolving the evolutionary history of a range of lineages, Asia has garnered a lot of attention (Gower et al., 2012). Many new species and larger lineages, as well as relicts and extinct taxa, can be found in this continental. Particularly, many different lineages have originated, propagated, and become extinct throughout China. China's topography and hydrography are diverse, with the tallest mountains in the world, several major rivers, salty lakes, and vast karst landscapes. Dating back to 1848 through 2019, 1,128 taxa have been discovered from China, with 421 taxa identified between 2000 and 2019, including six new genera: *Eunotiales*, *Bacillariales*, *Naviculales*, *Cymbellales*, *Rhopalodioid*, and *Surirellales* (Kociolek et al., 2020).

Lately, *Chamaepinnularia salina*, a new *Chamaepinnularia* species (Bacillariophyceae) was discovered at the mineral springs in the French Massif Central region. The valve shape of this species is linear with nearly parallel margins, and the longest individuals are slightly swollen in the middle, whereas the striae are made up of two big areolae, on the valve face and the mantle, respectively. As compared to *C. gandrupii*, *C. krasskei*, and *C. minutissima*, they have an elliptical

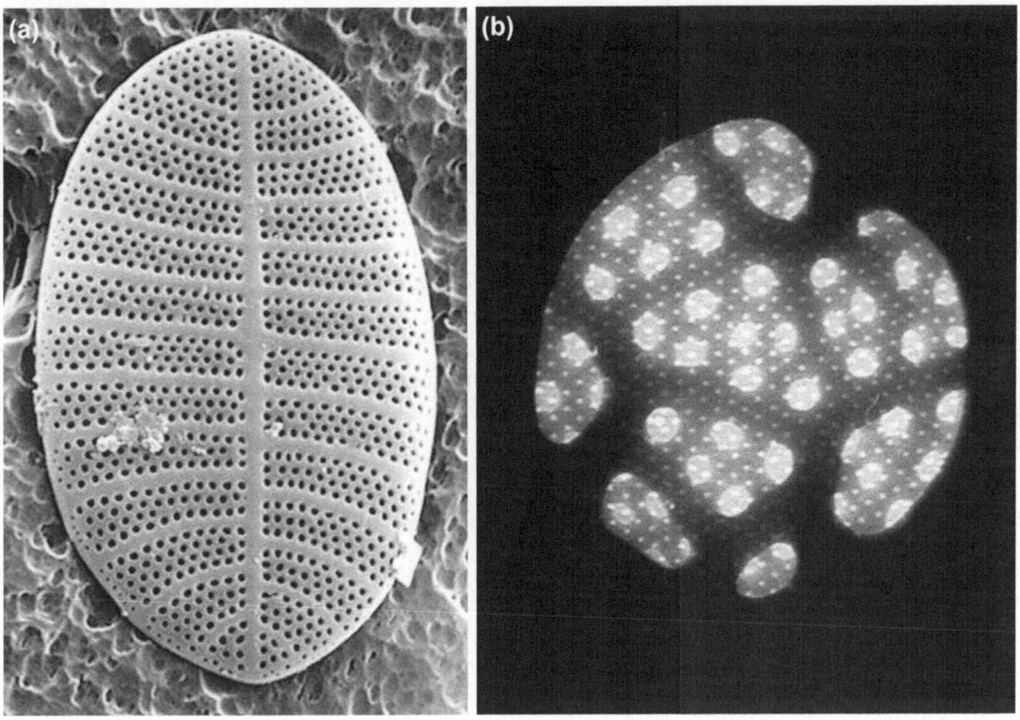

FIGURE 1.5 (a) The scanning electron micrograph representation of a monoraphid pennate (Cocconeis) diatom frustule; (b) Nanoscale silica biopatterning on the diatom (Lopez et al., 2005).

and rhombic-lanceolate central area shape, as well as *C. gerlachei* and *C. rupicola,* which are rhomboid to rectangular. *C. salina* has a rectangular central area, resulting in a slightly broad fascia. Besides, other unique structures of *C. salina* are the straight and fillform raphe, which is positioned on an elevated sternum. This species also has a higher striae density as compared to other similar species, which is 30–40 striae per 10 µm. According to the investigation, this species is mostly found in mineral springs containing Na-K-HCO_3 water type, which has high conductivity (Beauger et al., 2022).

In addition, scientists have also discovered two new and rare diatom species in the coastal water of Hawaii, which are the Rhopalodiaceae diatoms *Epithemia pelagica* and *Epithemia catenate*. These two distinct marine pennate diatom-diazotroph symbioses were studied by Schvarcz et al. (2022), where the diatoms were isolated from the subtropical North Pacific Ocean. This new marine phytoplankton species was claimed to become a self-fertilizing system by establishing a symbiotic relationship with nitrogen-fixing cyanobacteria. In the nutrient-deficit ocean waters of Hawaii, diatoms are unable to obtain adequate nitrogen to grow. As a solution, the symbiotic association with nitrogen-fixing cyanobacteria helps where cyanobacteria convert dissolved nitrogen gas into ammonia, a nitrogen source that diatoms can readily consume but is scarce in the open sea. Therefore, diatoms are able to alleviate the nitrogen-deficiency problem by having their own nitrogen generators (National Science Foundation, 2022; Schvarcz et al., 2022).

Furthermore, *Gomphosphenia vallei* (Bacillariophyceae) was discovered recently from a stream in the Réserve Naturelle Nationale de la Vallée de Chaudefour, Massif Central. It was identified from stone surfaces in Pérouse Stream. This new species is longer than *G. fontinalis*, in terms of its striae and composed of single slit-like areolae. It is characterized by a wide central area, which is absent to be asymmetrical and presented as a unilateral fascia. The short areola is only present on one side, and the axial area is not distinctly separated from the central area, along with a linear and narrow structure.

Internally, the proximal raphe endings are deflected in the same direction, and the frustules are almost rectangular to slightly wedge-shaped when observed in girdle view. On top of that, *G. vallei* is mostly found in running waters of low conductivity in a granitic and volcanic context, whereas other species are usually found in lakes, carbonate springs, and rivers (Aude et al., 2022).

Moreover, there are many other species discovered in different regions throughout the world. For instance, *Cymbopleura amicula* was discovered in Krka River, a groundwater-fed karst river in Croatia (Gligora et al., 2022); *Cocconeis czarneckii* was found in Lake Okoboji in the United States (Mora et al., 2022); discovery of *Melosira capsularum* was in the Jinsha River in China (Yang et al., 2022); *Eunotia crassiminor*, *Planothidium angustilanceolatum,* and *Delicatophycus crassiminutus* were discovered in spring habitats in the Northern Apennines in Italy (Cantonati et al., 2021), and many more. Undoubtedly, there are still many areas across the globe that await initial or additional in-depth investigation to reveal the undiscovered diatom species. More studies and research are needed as well to have a deeper understanding in terms of the potential application or contribution to the ecology of particular diatoms.

1.7 CHALLENGES AND FUTURE PERSPECTIVES

Diatoms can produce valuable bioactive molecules that can be used for paleontology, environmental and ecological studies, forensic limnology, pharmaceutical, food and beverages, etc. Despite the advantages of diatoms in different industries, there are some difficulties or issues faced for the commercialization of diatoms, particularly in the analysis of diatoms. This is due to the rigid and indestructible structure of the cell walls or frustules of diatoms. The conventional method to identify diatoms is based on the shape and structure of the valves, frustule characteristics, or other morphometric measurements of diatoms through the microscopy methods: electron microscopy (scanning electron microscopy) and light microscopy (Piper & Piper, 2020). The identification of diatoms using the microscopy method is time-consuming and laborious. Sample preparation with the treatment steps also can damage the structure of the diatom. Moreover, due to the large number of diatom species, species similarities, intraspecific diversity, and life-cycle variation, the manual identification of diatom species is difficult and challenging, requiring expertise or skilled taxonomists to prevent misidentification (Blanco et al., 2017; Caine, 2021). In addition, skilled taxonomists may reach different results when all the requisite features for identification are present, either as a result of various species delimitation criteria or different interpretations (Eduardo et al., 2001; Mann et al., 2010).

A study by Cox (2014) revealed that the changes in diatoms' morphology due to intrinsic and extrinsic changes will influence the findings of molecular, reproductive, and ecological use of diatoms, affecting the interpretation of the identity of diatoms. Intrinsic changes refer to changes in the form or morphology of diatoms due to the complex life cycle of the diatom that consists of asexual (cell division) and sexual reproductive stages. The morphological features and characteristics are diversified within the population or clonal cultures of diatoms (Bart Van De et al., 2013; Zinger et al., 2012). Throughout the life cycle, the diatoms will experience changes in size, shape, gridle, and valves (Blanco et al., 2017; Cox, 2014; Sánchez et al., 2019b). Accordingly, identifying diatoms morphologically requires expertise in taxonomic knowledge and advanced high-resolution microscopy tools. Obtaining the sufficient number of images of each species of diatom at each stage of its life cycle is a difficult task. Next, the environmental conditions such as temperature, light, pH, salinity, and presence of heavy metals or toxicants in the environment can induce the changes in the frustules and morphology of diatoms and thus affect the identification of diatoms. Godhe and Rynearson (2017) showed that environmental factors can cause intraspecific variation in diatoms for survival. Cortese and Gersonde's (2007) study in the Southern Ocean showed that the valve size of *Fragilariopsis kerguelensis* was reduced from glacial intervals to interglacial counterparts. In short, the diverse variability in characteristics, morphological, and morphometric features of diatoms is due to intrinsic and extrinsic factors that must be considered during the identification procedure (Cristóbal et al., 2020).

Therefore, the prospect of the diatom identification process has improved remarkably in synch with the advancement in microscopy and metabarcoding approaches as well as incorporation with artificial intelligence (Zhou et al., 2020). The concept behind is image acquisition by microscopy approach, segmentation of the images, followed by taxonomic identification by artificial intelligence models. Zimmermann et al. (2015) compared the two identification approaches of diatoms from environmental samples, which were metabarcoding and light microscopy. The finding revealed that the next generation sequencing eDNA barcoding approach and light microscopy demonstrated similar results, but metabarcoding can provide a greater taxonomic depth. Kloster et al. (2020) proposed a diatom identification method by combining a high-resolution microscopy technique with convolutional neural network-based classification integrated with web-based taxonomic annotation of gigapixel-sized "virtual slides". In May 1998, there was one pilot project named ADIAC (Automatic Diatom Identification and Classification) initiated by the European MAST (Marine Science and Technology) program to create algorithms to identify diatoms automatically through the utilization of image processing and pattern recognition tools, including valve shape (contour) and ornamentation because many diatoms cannot be differentiated only on the basis of morphology since different species have similar shapes. In addition, this project introduced other essential features such as automatic slide scanning, autofocusing, production of large image libraries, and integration into taxonomic or ecological database systems. As a result, this project attempted to develop a software for diatom slide scanning, autofocusing, image analysis, and pattern identification. Large image databases, as well as taxonomic and ecological descriptions, were to be generated simultaneously (Du Buf & Bayer, 2002). In summary, an artificial intelligence system might potentially replace the task of manually observing the morphology of diatoms throughout the diatom identification process.

1.8 CONCLUSION

A diatom is a photosynthetic organism with a single cell that indicates it will generate foodstuff by itself in the identical manner as plants perform. Most aquatic ecosystems' primary and secondary consumers rely on diatoms to some extent and even completely. Over the years, with the advent of microscopes, the studies related to diatoms have progressed rapidly, and more researchers have switched their interest of study in investigating the details of morphological features of diatoms. Various investigations have also been accomplished to compare the diatom identification methods for environmental assessments between morphological and DNA metabarcoding. Undoubtedly, there are still many areas across the globe that await initial or additional in-depth investigation to reveal the undiscovered diatom species. More studies and research are needed as well to have a deeper understanding in terms of the potential application or contribution to the ecology of particular diatoms.

REFERENCES

Apothéloz-Perret-Gentil, L., Bouchez, A., Cordier, T., Cordonier, A., Guéguen, J., Rimet, F., …Pawlowski, J. (2021). Monitoring the ecological status of rivers with diatom eDNA metabarcoding: A comparison of taxonomic markers and analytical approaches for the inference of a molecular diatom index. *Molecular Ecology*, *30*(13), 2959–2968.

Arini, A., Durant, F., Coste, M., Delmas, F., & Feurtet-Mazel, A. (2013). Cadmium decontamination and reversal potential of teratological forms of the diatom Planothidium frequentissimum (Bacillariophyceae) after experimental contamination. *Journal of Phycology*, *49*(2), 361–370.

Armbrust, E. V. (2009). The life of diatoms in the world's oceans. *Nature*, *459*(7244), 185–192. doi:10.1038/nature08057

Aude, B., Elisabeth, A., Olivier, V., Christelle, B., Sofia, R., Carlos, E. W., & Luc, E. (2022). Gomphosphenia vallei (Bacillariophyta), a new diatom species from a stream in the "Réserve Naturelle Nationale de la Vallée de Chaudefour", Massif Central (France). *Phytotaxa*, *542*(2), 167–179-167–179.

Barral-Fraga, L., Morin, S., Rovira, M. D., Urrea, G., Magellan, K., & Guasch, H. (2016). Short-term arsenic exposure reduces diatom cell size in biofilm communities. *Environmental Science and Pollution Research*, *23*(5), 4257–4270.

Bart Van De, V., Amelie, J., Myriam De, H., & Pierre, C. (2013). The Ultrastructure of Gomphonema augur and Gomphonema gautieri (Bacillariophyta). *Cryptogamie, Algologie*, *34*(2), 103–116. doi:10.7872/crya.v34.iss2.2013.103

Bayer, M., & Mann, D. (2000). The Diatoms: Applications for the environmental and earth sciences. Edited by Eugene F. Stoermer & John P. Smol. *Edinburgh Journal of Botany*, *57*, 141–150. doi:10.1017/S0960428600210093

Beauger, A., Wetzel, C., Allain, E., Bertin, C., Voldoire, O., Breton, V., …Ector, L. (2022). Chamaepinnularia salina (Bacillariophyta), a new diatom species from French mineral springs (Massif Central). *Phytotaxa*, *538*(1), 55–73.

Bidgood, M., Mitlehner, A., Jones, G., & Jutson, D. (1999). Towards a stable and agreed nomenclature for North Sea Tertiary diatom floras – The 'Coscinodiscus' problem. *Geological Society, London, Special Publications*, *152*, 139–153. doi:10.1144/GSL.SP.1999.152.01.08

Blanco, S. (2020). Diatom taxonomy and identification keys. *Modern Trends in Diatom Identification*, 10, Springer. 25–38.

Blanco, S., Borrego-Ramos, M., & Olenici, A. (2017). Disentangling diatom species complexes: Does morphometry suffice? *PeerJ*, *5*, e4159.

Caine, J. (2021). Diatom analysis: Reviewing the strengths, weaknesses, and impacts of modern research: Exploration of the Science, Strengths, Weaknesses, and the Future. *Canadian Journal of Undergraduate Research*, *6*(1).

Cairns, J., Dickson, K., & Slocomb, J. (1977). The ABC's of diatom identification using laser holography. *Hydrobiologia*, *54*(1), 7–16.

Calvo-Flores, M. D., Contreras, W. F., Galindo, E. G., & Pérez-Pérez, R. (2006). XKey: A tool for the generation of identification keys. *Expert Systems with Applications*, *30*(2), 337–351.

Cantonati, M., Bilous, O., Angeli, N., van Wensen, L., & Lange-Bertalot, H. (2021). Three new diatom species from spring habitats in the northern Apennines (Emilia-Romagna, Italy). *Diversity*, *13*(11), 549.

Christensen, T. (1991). The diatoms. Biology and morphology of the genera: Taylor & Francis.

Clark, J. Y., & Warwick, K. (1998). Artificial keys for botanical identification using a multilayer perceptron neural network (MLP). Panigrahi S., Ting K. C. (eds) *Artificial intelligence for biology and agriculture* (pp. 95–115): Springer.

Coquillé, N., & Morin, S. (2019). Fitness of teratological morphotypes and heritability of deformities in the diatom Gomphonema gracile. *Ecological Indicators*, *106*, 105442.

Cortese, G., & Gersonde, R. (2007). Morphometric variability in the diatom Fragilariopsis kerguelensis: Implications for Southern Ocean paleoceanography. *Earth and Planetary Science Letters*, *257*(3), 526–544. doi:10.1016/j.epsl.2007.03.021

Cox, E. J. (2010). Morphogenetic information and the selection of taxonomic characters for raphid diatom systematics. *Plant Ecology and Evolution*, *143*(3), 271–277.

Cox, E. J. (2014). Diatom identification in the face of changing species concepts and evidence of phenotypic plasticity. *J. Micropalaeontol.*, *33*(2), 111–120. doi:10.1144/jmpaleo2014-014

Cox, E. J. (2015).Coscinodiscophyceae, Mediophyceae, Fragilariophyceae, Bacilariopphyceae (Diatoms). Frey W (ed) *Syllabus of Plant Families. Adolf Engler's Syllabus der Pflanzenfamilien* (pp. 64–103): Borntraeger Science Publishers.

Cristóbal, G., Blanco, S., & Bueno, G. (2020). *Modern trends in diatom identification*: Springer.

da Silva, E. F., Almeida, S. F., Nunes, M. L., Luís, A. T., Borg, F., Hedlund, M., …Teixeira, P. (2009). Heavy metal pollution downstream the abandoned Coval da Mó mine (Portugal) and associated effects on epilithic diatom communities. *Science of the Total Environment*, *407*(21), 5620–5636.

Dimitrovski, I., Kocev, D., Loskovska, S., & Džeroski, S. (2012). Hierarchical classification of diatom images using ensembles of predictive clustering trees. *Ecological Informatics*, *7*(1), 19–29. doi:10.1016/j.ecoinf.2011.09.001

Du Buf, H., & Bayer, M. (2002). Series in machine perception and artificial intelligence. *Automatic diatom identification*: World Scientific Publishing.

Edlund, M. B., & Stoermer, E. F. (1997). Ecological, evolutionary, and systematic significance of diatom life histories 1. *Journal of Phycology*, *33*(6), 897–918. doi: 10.1111/j.0022-3646.1997.00897.x

Dunn, G., & Everitt, B. S. (2004). *An introduction to mathematical taxonomy*: Courier Corporation.

Eduardo, A. M., Peter, A. S., & Francis, R. T. (2001). Identification of diatoms (Bacillariophyceae) during ecological assessments: Comparison between Light Microscopy and Scanning Electron Microscopy techniques. *Proceedings of the Academy of Natural Sciences of Philadelphia, 151*(1), 95–103. doi: 10.1635/0097-3157(2001)151[0095:IODBDE]2.0.CO;2

Evans, K. M., & Mann, D. G. (2009). A proposed protocol for nomenclaturally effective DNA barcoding of microalgae. *Phycologia, 48*(1), 70–74.

Falasco, E., Bona, F., Ginepro, M., Hlúbiková, D., Hoffmann, L., & Ector, L. (2009). Morphological abnormalities of diatom silica walls in relation to heavy metal contamination and artificial growth conditions. *Water sa, 35*(5), 595-606.

Falasco, E., Ector, L., Wetzel, C. E., Badino, G., & Bona, F. (2021). Looking back, looking forward: a review of the new literature on diatom teratological forms (2010–2020). *Hydrobiologia, 848*(8), 1675–1753.

Falasco, E., Piano, E., Doretto, A., Fenoglio, S., & Bona, F. (2018). Resilience of benthic diatom communities in Mediterranean streams: Role of endangered species. *Marine and Freshwater Research, 70*(2), 212–224.

Feio, M. J., Serra, S. R., Mortágua, A., Bouchez, A., Rimet, F., Vasselon, V., & Almeida, S. F. (2020). A taxonomy-free approach based on machine learning to assess the quality of rivers with diatoms. *Science of the Total Environment, 722*, 137900.

Finkel, Z., Katz, M., Wright, J., Schofield, O., & Falkowski, P. (2005). Climatically driven macroevolutionary patterns in the size of marine diatoms of the Cenozoic. *Proceedings of the National Academy of Sciences of the United States of America, 102*, 8927–8932. doi:10.1073/pnas.0409907102

Finkel, Z. V., Beardall, J., Flynn, K. J., Quigg, A., Rees, T. A. V., & Raven, J. A. (2010). Phytoplankton in a changing world: Cell size and elemental stoichiometry. *Journal of plankton research, 32*(1), 119–137.

Forio, M. A. E., & Goethals, P. L. (2020). An integrated approach of multi-community monitoring and assessment of aquatic ecosystems to support sustainable development. *Sustainability, 12*(14), 5603.

Fu, W., Shu, Y., Yi, Z., Su, Y., Pan, Y., Zhang, F., & Brynjolfsson, S. (2022). Diatom morphology and adaptation: Current progress and potentials for sustainable development. *Sustainable Horizons, 2*, 100015.

Gligora, U. M., Antonija, K., Mirela, Š., Jasna, A., Saul, B., & Zlatko, L. (2022). Cymbopleura amicula stat nov. et. nom. nov.(Bacillariophyceae)—A rare diatom species from a karst river in Croatia. *Phytotaxa, 532*(2), 139–151.

Godhe, A., Kremp, A., & Montresor, M. (2014). Genetic and microscopic evidence for sexual reproduction in the centric diatom Skeletonema marinoi. *Protist, 165*(4), 401–416. doi:10.1016/j.protis.2014.04.006

Godhe, A., & Rynearson, T. (2017). The role of intraspecific variation in the ecological and evolutionary success of diatoms in changing environments. *Philosophical Transactions of the Royal Society B: Biological Sciences, 372*(1728), 20160399.

Gower, D., Johnson, K., Richardson, J., Rosen, B., Rüber, L., & Williams, S. (2012). *Biotic evolution and environmental change in southeast Asia* (Vol. 82): Cambridge University Press.

Hall, R. I., & Smol, J. P. (2010). Diatoms as indicators of lake eutrophication. In E. F. Stoermer & J. P. Smol (Eds.), *The diatoms: Applications for the environmental and earth sciences* (2nd ed., pp. 122–151): Cambridge University Press.

Hamm, C. E., Merkel, R., Springer, O., Jurkojc, P., Maier, C., Prechtel, K., & Smetacek, V. (2003). Architecture and material properties of diatom shells provide effective mechanical protection. *Nature, 421*(6925), 841–843. doi:10.1038/nature01416

Hamsher, S. E., Evans, K. M., Mann, D. G., Poulíčková, A., & Saunders, G. W. (2011). Barcoding diatoms: Exploring alternatives to COI-5P. *Protist, 162*(3), 405–422.

Hendey, N. I. (1951). Littoral diatoms op Chichester harbour with special reference to fouling. *Journal of the Royal Microscopical Society, 71*(1), 1–86.

Hinder, S. L., Hays, G. C., Edwards, M., Roberts, E. C., Walne, A. W., & Gravenor, M. B. (2012). Changes in marine dinoflagellate and diatom abundance under climate change. *Nature Climate Change, 2*(4), 271–275. doi:10.1038/nclimate1388

Hu, Y., Kumar, M., Wang, Z., Zhan, X., & Stengel, D. B. (2021). Filamentous microalgae as an advantageous co-substrate for enhanced methane production and digestate dewaterability in anaerobic co-digestion of pig manure. *Waste Management, 119*, 399–407.

Jackson, T., Bouman, H., Sathyendranath, S., & Devred, E. (2010). Regional-scale changes in diatom distribution in the Humboldt upwelling system as revealed by remote sensing: Implications for fisheries. *ICES Journal of Marine Science, 68*, 729–736. doi:10.1093/icesjms/fsq181

Joseph, S., & Patrick, K. (2011). The Diatom World: Cellular Origin, Life in Extreme Habitats and Astrobiology, Springer Dordrecht. doi:10.1007/978-94-007-1327-7

Kale, A., & Karthick, B. (2015). The diatoms. *Resonance*, *20*(10), 919–930.

Karbalaie, Z., Balali, G. R., Akhavan Roofigar, A., & Bagheri, A. (2021). A numerical taxonomic study of the genus Artemisia L.(Asteraceae) in Isfahan Province. *Journal of Taxonomy and Biosystematics*, *47*, 1–14.

Kloster, M., Langenkämper, D., Zurowietz, M., Beszteri, B., & Nattkemper, T. W. (2020). Deep learning-based diatom taxonomy on virtual slides. *Scientific Reports*, *10*(1), 14416. doi:10.1038/s41598-020-71165-w

Kociolek, J. P., & Williams, D. M. (1987). Unicell ontogeny and phylogeny: Examples from the diatoms. Cladistics, 3, 274-284. doi: 10.1111/j.1096-0031.1987.tb00511.x

Kociolek, J. P., You, Q., Liu, Q., Liu, Y., & Wang, Q. (2020). Continental diatom biodiversity discovery and description in China: 1848 through 2019. *PhytoKeys*, *160*, 45.

Kooistra, W. H., Sarno, D., Balzano, S., Gu, H., Andersen, R. A., & Zingone, A. (2008). Global diversity and biogeography of Skeletonema species (Bacillariophyta). *Protist*, *159*(2), 177–193.

Kooistra, W. H. C. F., Gersonde, R., Medlin, L. K., & Mann, D. G. (2007). The origin and evolution of the diatoms: Their adaptation to a planktonic existence. In P. G. Falkowski & A. H. Knoll (Eds.), *Evolution of primary producers in the sea* (pp. 207–249): Academic Press.

Koreneva, T., & Latkovskaya, E. (2013). Teratological forms of diatomic algae in alga-flora of the Lake Teletskoye (Russian Altai). *Water Chem. Ecol.*, *10*, 112–119.

Kulkarni, R. (2020). Tiny but mighty: Newly-discovered diatom from Meghalaya's cave puts the spotlight on these algae. Retrieved from https://researchmatters.in/news/tiny-mighty-newly-discovered-diatom-meghalaya%E2%80%99s-cave-puts-spotlight-these-algae#:~:text=In%20recent%20years%2C%20many%20species,Cherrapunji%2C%20where%20it%20was%20found.

Kumaran, J., Jose, B., Joseph, V., & Singh, B. (2016). Optimization of growth requirements of marine diatom Chaetoceros muelleri using Response Surface Methodology. *Aquaculture Research*, *48*, 1513-1524. doi:10.1111/are.12987

Leguay, S., Lavoie, I., Levy, J. L., & Fortin, C. (2016). Using biofilms for monitoring metal contamination in lotic ecosystems: The protective effects of hardness and pH on metal bioaccumulation. *Environmental Toxicology and Chemistry*, *35*(6), 1489–1501.

Lei, Y., Zhou, X., & Xie, L. (2019). Emergency monitoring and disposal decision support system for sudden pollution accidents based on multimedia information system. *Multimedia Tools and Applications*, *78*(8), 11047–11071.

Leuschner, D., & Sviridov, A. (1986). The mathematical theory of taxonomic keys. *Biometrical Journal*, *28*(1), 109–113.

Li, Y.-L., Chen, X., Xiao, X., Zhang, H., Xue, B., Shen, J., & Zhang, E. (2018). Diatom-based inference of Asian monsoon precipitation from a volcanic lake in southwest China for the last 18.5 ka. *Quaternary Science Reviews*, *182*, 109–120. doi:10.1016/j.quascirev.2017.11.021

Loke, R. E., du Buf, J. M. H., Bayer, M. M., & Mann, D. G. (2004). Diatom classification in ecological applications. *Pattern Recognition*, *37*(6), 1283–1285. doi:10.1016/j.patcog.2003.08.018

Lopez, P. J., Desclés, J., Allen, A. E., & Bowler, C. (2005). Prospects in diatom research. *Current Opinion in Biotechnology*, *16*(2), 180–186. doi:10.1016/j.copbio.2005.02.002

Luís, A. T., Teixeira, P., Almeida, S. F. P., Matos, J. X., & da Silva, E. F. (2011). Environmental impact of mining activities in the Lousal area (Portugal): Chemical and diatom characterization of metal-contaminated stream sediments and surface water of Corona stream. *Science of the Total Environment*, *409*(20), 4312–4325.

Lund, J. W. G. (1946). The ecology, size and taxonomy of British soil diatoms. Part II. *New Phytol*, *45*, 56-110.

Lundholm, N., Krock, B., John, U., Skov, J., Cheng, J., Pančić, M., …Harðardóttir, S. (2018). Induction of domoic acid production in diatoms—Types of grazers and diatoms are important. *Harmful Algae*, *79*, 64–73. doi:10.1016/j.hal.2018.06.005

Majewska, R., Zgrundo, A., Lemke, P., & De Stefano, M. (2012). Benthic diatoms of the Vistula River estuary (Northern Poland): Seasonality, substrata preferences, and the influence of water chemistry. *Phycological Research*, *60*(1), 1–19.

Mancuso, J. L., Weinke, A. D., Stone, I. P., Hamsher, S. E., Villar-Argaiz, M., & Biddanda, B. A. (2021). Cold and wet: Diatoms dominate the phytoplankton community during a year of anomalous weather in a Great Lakes estuary. *Journal of Great Lakes Research*, *47*(5), 1305–1315.

Mann, D. G. (2010). Discovering diatom species: is a long history of disagreements about species-level taxonomy now at an end? *Plant Ecology and Evolution*, *143*(3), 251–264.

Mann, D. G. (2011). Size and sex. In J. Seckbach & P. Kociolek (Eds.), *The diatom world* (pp. 145–166): Springer Netherlands.

Mann, D. G., & Droop, S. (1996). Biodiversity, biogeography and conservation of diatoms. Kristiansen, J. (eds). *Biogeography of freshwater algae* (pp. 19–32): Springer.

Mann, D. G., Sato, S., Trobajo, R., Vanormelingen, P., & Souffreau, C. (2010). DNA barcoding for species identification and discovery in diatoms. *Cryptogamie Algologie*, *31*, 557–577.

Masmoudi, S., Nguyen-Deroche, N., Caruso, A., Ayadi, H., Morant-Manceau, A., & Tremblin, G. (2013). Cadmium, copper, sodium and zinc effects on diatoms: From heaven to hell—A review. *Cryptogamie, Algologie*, *34*(2), 185–225.

Meyer, B., Wulf, M., & Håkansson, H. (2001). Phenotypic variation of life-cycle stages in clones of three similar Cyclotella species after induced auxospore production. *Diatom Research*, *16*(2), 343–361.

Moeys, S., Frenkel, J., Lembke, C., Gillard, J. T. F., Devos, V., Van den Berge, K., …Vyverman, W. (2016). A sex-inducing pheromone triggers cell cycle arrest and mate attraction in the diatom Seminavis robusta. *Scientific Reports*, *6*(1), 19252. doi:10.1038/srep19252

Moore, E. R., Bullington, B. S., Weisberg, A. J., Jiang, Y., Chang, J., & Halsey, K. H. (2017). Morphological and transcriptomic evidence for ammonium induction of sexual reproduction in Thalassiosira pseudonana and other centric diatoms. *PLOS One*, *12*(7), e0181098. doi:10.1371/journal.pone.0181098

Mora, D., Abarca, N., Proft, S., Grau, J. H., Enke, N., Carmona, J., …Zimmermann, J. (2019). Morphology and metabarcoding: A test with stream diatoms from Mexico highlights the complementarity of identification methods. *Freshwater Science*, *38*(3), 448–464.

Mora, D., Stancheva, R., & Jahn, R. (2022). Cocconeis czarneckii sp. nov.(Bacillariophyta): A new diatom species from Lake Okoboji (Iowa, USA), based on the strain UTEX FD23. *Phycologia*, *61*(1), 60–74.

Mortágua, A., Vasselon, V., Oliveira, R., Elias, C., Chardon, C., Bouchez, A., …Almeida, S. F. (2019). Applicability of DNA metabarcoding approach in the bioassessment of Portuguese rivers using diatoms. *Ecological Indicators*, *106*, 105470.

Munnecke, A., Delabroye, A., Servais, T., Vandenbroucke, T. R., & Vecoli, M. (2012). Systematic occurrences of malformed (teratological) acritarchs in the run-up of Early Palaeozoic δ13C isotope excursions. *Palaeogeography, Palaeoclimatology, Palaeoecology*, *367*, 137–146.

National Science Foundation. (2022). Scientists discover new and unusual species of diatoms in waters off Hawaii. Retrieved from https://www.nsf.gov/discoveries/disc_summ.jsp?cntn_id=304662&org=NSF

Olenici, A., Baciu, C., Blanco, S., & Morin, S. (2020). Naturally and environmentally driven variations in diatom morphology: Implications for diatom-based assessment of water quality. Cristóbal G. Blanco S. Bueno G. (eds) *Modern trends in diatom identification* (pp. 39–50): Springer.

Olenici, A., Blanco, S., Borrego-Ramos, M., Jiménez-Gómez, F., Guerrero, F., Momeu, L., & Baciu, C. (2019). A new diatom teratology driven by metal pollution in a temperate river (Roșia Montană, Romania). *Annali di Botanica*, *9*, 113–118.

Ouyang, L.-l., Pan, Y.-d., Huang, C.-m., Tang, Y., Du, J., & Xiao, W.-y. (2016). Water quality assessment of benthic diatom communities for water quality in the subalpine karstic lakes of Jiuzhaigou, a world heritage site in China. *Journal of Mountain Science*, *13*(9), 1632–1644.

Pandey, L. K., & Bergey, E. A. (2016). Exploring the status of motility, lipid bodies, deformities and size reduction in periphytic diatom community from chronically metal (Cu, Zn) polluted waterbodies as a biomonitoring tool. *Science of the Total Environment*, *550*, 372–381.

Pandey, L. K., Bergey, E. A., Lyu, J., Park, J., Choi, S., Lee, H., …Han, T. (2017). The use of diatoms in ecotoxicology and bioassessment: Insights, advances and challenges. *Water Research*, *118*, 39–58.

Pappas, J., Kociolek, P., & Stoermer, E. F. (2014). Quantitative morphometric methods in diatom research. *Nova Hedwig*, *143*, 281–306.

Pappas, J. L., & Stoermer, E. F. (2003). Morphometric comparison of the neotype of Asterionella formosa Hassall (Heterokontophyta, Bacillariophyceae) with Asterionella edlundii s p. nov. from Lake Hovsgol, Mongolia. *Diatom*, *19*, 55–65.

Passy, S. I. (2007). Differential cell size optimization strategies produce distinct diatom richness–body size relationships in stream benthos and plankton. *Journal of Ecology*, *95*(4), 745–754.

Payne, R., & Preece, D. (1980). Identification keys and diagnostic tables: A review. *Journal of the Royal Statistical Society: Series A (General)*, *143*(3), 253–282.

Pfister, C. A., & Stevens, F. R. (2002). The genesis of size variability in plants and animals. *Ecology*, *83*(1), 59–72.

Pierella Karlusich, J. J., Bowler, C., & Biswas, H. (2021). Carbon dioxide concentration mechanisms in natural populations of marine diatoms: Insights from Tara oceans. *Frontiers in Plant Science*, *12*, 657821. doi:10.3389/fpls.2021.657821

Piper, J., & Piper, T. (2020). Microscopic modalities and illumination techniques. In G. Cristóbal, S. Blanco, & G. Bueno (Eds.), *Modern trends in diatom identification: Fundamentals and applications* (pp. 53–93): Springer International Publishing.

Potapova, M., & Hamilton, P. B. (2007). Morphological and ecological variation within the Achnanthidium minutissimum (Bacillariophyceae) species complex 1. *Journal of Phycology*, *43*(3), 561–575.

Poulíčková, A., & Mann, D. G. (2008). Autogamous auxosporulation in Pinnularia nodosa (Bacillariophyceae)1. *Journal of Phycology*, *44*(2), 350–363. doi:10.1111/j.1529-8817.2008.00475.x

Reynolds, A. P., Dicks, J. L., Roberts, I. N., Wesselink, J.-J., Iglesia, B. d. l., Robert, V., …Rayward-Smith, V. J. (2003). *Algorithms for identification key generation and optimization with application to yeast identification*. Paper presented at the Workshops on Applications of Evolutionary Computation.

Rivera, S., Vasselon, V., Jacquet, S., Bouchez, A., Ariztegui, D., & Rimet, F. (2018). Metabarcoding of lake benthic diatoms: from structure assemblages to ecological assessment. *Hydrobiologia*, *807*(1), 37–51.

Roy, S., & Karthick, B. (2019). Diatoms from the Congo and Zambezi Basins – Methodologies and identification of the genera. *African Journal of Aquatic Science*, *44*(3), 301–302. doi:10.2989/16085914.2019.1628702

Rühland, K. M., Paterson, A. M., & Smol, J. P. (2015). Lake diatom responses to warming: Reviewing the evidence. *Journal of Paleolimnology*, *54*(1), 1–35. doi:10.1007/s10933-015-9837-3

Saade, A., & Bowler, C. (2009). Molecular tools for discovering the secrets of diatoms. *BioScience*, *59*(9), 757–765. doi:10.1525/bio.2009.59.9.7

Salido, J., Sánchez, C., Ruiz-Santaquiteria, J., Cristóbal, G., Blanco, S., & Bueno, G. (2020). A low-cost automated digital microscopy platform for automatic identification of diatoms. *Applied Sciences*, *10*(17), 6033.

Sánchez, C., Cristóbal, G., & Bueno, G. (2019a). Diatom identification including life cycle stages through morphological and texture descriptors. *PeerJ*, *7*, e6770–e6770. doi:10.7717/peerj.6770

Sánchez, C., Cristóbal, G., & Bueno, G. (2019b). Diatom identification including life cycle stages through morphological and texture descriptors. *PeerJ*, *7*, e6770.

Sato, S., Mann, D. G., Nagumo, T., Tanaka, J., Tadano, T., & Medlin, L. K. (2008). Auxospore fine structure and variation in modes of cell size changes in Grammatophora marina (Bacillariophyta). *Phycologia*, *47*(1), 12–27. doi:10.2216/07-34.1

Schmid, A.-M. M. (1994). Aspects of morphogenesis and function of diatom cell walls with implications for taxonomy. *Protoplasma* 181, 43–60. doi:10.1007/BF01666388

Schvarcz, C. R., Wilson, S. T., Caffin, M., Stancheva, R., Li, Q., Turk-Kubo, K. A., …Steward, G. F. (2022). Overlooked and widespread pennate diatom-diazotroph symbioses in the sea. *Nature Communications*, *13*(1), 1–9.

Seckbach, J., & Kociolek, P. (2011). *The diatom world* (Vol. 19): Springer Science & Business Media.

Sharma, N., Simon, D. P., Diaz-Garza, A. M., Fantino, E., Messaabi, A., Meddeb-Mouelhi, F., …Desgagné-Penix, I. (2021). Diatoms biotechnology: Various industrial applications for a greener tomorrow. *Frontiers in Marine Science*, *8*, 636613. doi:10.3389/fmars.2021.636613

Shen, R., Ren, H., Yu, P., You, Q., Pang, W., & Wang, Q. (2018). Benthic diatoms of the Ying River (Huaihe River Basin, China) and their application in water trophic status assessment. *Water*, *10*(8), 1013.

Shukla, S. K., & Romero, O. E. (2018). Glacial valve size variation of the Southern Ocean diatom Fragilariopsis kerguelensis preserved in the Benguela Upwelling System, southeastern Atlantic. *Palaeogeography, Palaeoclimatology, Palaeoecology*, *499*, 112–122.

Talent, N., Dickinson, R. B., & Dickinson, T. A. (2014). Character selection during interactive taxonomic identification:"Best Characters". *Biodiversity Informatics*, *9*, 1–12.

Tapolczai, K., Selmeczy, G. B., Szabó, B., Viktória, B., Keck, F., Bouchez, A., …Padisák, J. (2021). The potential of exact sequence variants (ESVs) to interpret and assess the impact of agricultural pressure on stream diatom assemblages revealed by DNA metabarcoding. *Ecological Indicators*, *122*, 107322.

Tréguer, P., Bowler, C., Moriceau, B., Dutkiewicz, S., Gehlen, M., Aumont, O., …Pondaven, P. (2018). Influence of diatom diversity on the ocean biological carbon pump. *Nature Geoscience*, *11*(1), 27–37. doi:10.1038/s41561-017-0028-x

Tréguer, P., & Rocha, C. (2013). The world ocean silica cycle. *Annual Review of Marine Science*, *5*, 477–501. doi:10.1146/annurev-marine-121211-172346

Turland, N. J., Wiersema, J. H., Barrie, F. R., Greuter, W., Hawksworth, D. L., Herendeen, P. S., …Marhold, K. (2018). *International code of nomenclature for algae, fungi, and plants (Shenzhen Code) adopted by the Nineteenth International Botanical Congress Shenzhen, China, July 2017*: Koeltz Botanical Books.

Valentin, V., Frédéric, R., Isabelle, D., Olivier, M., Yorick, R., & Agnès, B. (2019). Assessing pollution of aquatic environments with diatoms' DNA metabarcoding: Experience and developments from France Water Framework Directive networks. *Metabarcoding and Metagenomics*, *3*, e39646.

Walter, D. E., & Winterton, S. (2007). Keys and the crisis in taxonomy: Extinction or reinvention? *Annu. Rev. Entomol.*, *52*, 193–208.

Wang, Q., Yang, X., Anderson, N., Zhang, E., & Li, Y.-L. (2014). Diatom response to climate forcing of a deep, alpine lake (Lugu Hu, Yunnan, SW China) during the Last Glacial Maximum and its implications for understanding regional monsoon variability. *Quaternary Science Reviews*, *86*, 1–12. doi:10.1016/j.quascirev.2013.12.024

Williams, D. (2020). Cristóbal G Blanco S Bueno G Diatom classifications: What purpose do they serve? (eds) Modern Trends in Diatom Identification. Developments in Applied Phycology, 10. (pp. 11–24). Springer.

Williams, D. M., & Kociolek, J. P. (2011). An overview of diatom classification with some prospects for the future. Seckbach, J., Kociolek, P. (eds) *The Diatom World.* Cellular Origin, Life in Extreme Habitats and Astrobiology, 19, 47–91, Springer.

Yang, L., Yu, P., You, Q., Li, G., & Wang, Q. (2022). Morphological and phylogenetic analysis of a new Melosira species and revision of freshwater Melosira in China. *Journal of Oceanology and Limnology*, *40*(2), 712–728.

Zhang, X.-b., Chen, X.-x., & Cheng, J.-a. (2008). SIKey: A tool to generate secondary identification keys for targeted diagnosis. *Expert Systems with Applications*, *34*(4), 2812–2818.

Zhang, Y., Tao, J., Wang, J., Ding, L., Ding, C., Li, Y., …Zhang, H. (2019). Trends in diatom research since 1991 based on topic modeling. *Microorganisms*, *7*(8), 213.

Zhang, Y. X. (2019). Current diatom research in China. Seckbach, J., Gordon, R. *Diatoms: Fundamentals and Applications*, 43–98. Scrivener.

Zhao, Y., Chen, X., Yang, Y., Zhao, X., Zhang, S., Gao, Z., …Zhang, J. (2018). Potential forensic biogeographic application of diatom colony consistency analysis employing pyrosequencing profiles of the 18S rDNA V7 region. *International Journal of Legal Medicine*, *132*(6), 1611–1620.

Zhou, Y., Cao, Y., Huang, J., Deng, K., Ma, K., Zhang, T., …Huang, P. (2020). Research advances in forensic diatom testing. *Forensic Sciences Research*, *5*(2), 98–105.

Zimmermann, J., Glöckner, G., Jahn, R., Enke, N., & Gemeinholzer, B. (2015). Metabarcoding vs. morphological identification to assess diatom diversity in environmental studies. *Molecular Ecology Resources*, *15*(3), 526–542.

Zinger, L., Gobet, A., & Pommier, T. (2012). Two decades of describing the unseen majority of aquatic microbial diversity. *Molecular Ecology*, *21*(8), 1878–1896. doi:10.1111/j.1365-294X.2011.05362.x

2 Diatom High Cell Density Culturing

Challenges and Prospects

Thomas Kiran Marella
Algae Biomass and Energy Systems R&D Center (ABES), University of Tsukuba, Tsukuba, Ibaraki, Japan

Meenakshi Singh
Department of Ecology & Biodiversity, Terracon Ecotech Pvt. Ltd., Mumbai, Maharashtra, India

K. Chandrasekhar
School of Civil and Environmental Engineering, Yonsei University, Seoul, Republic of Korea

Pankaj Kumar Singh and Rashi Tyagi
Diatom Research Laboratory, Amity Institute of Biotechnology, Amity University, Noida, Uttar Pradesh, India

Makoto M. Watanabe
Algae Biomass and Energy Systems R&D Center (ABES), University of Tsukuba, Tsukuba, Ibaraki, Japan

CONTENTS

2.1 INTRODUCTION

In recent years, tremendous research interest has been generated among microalgae researchers for their use as renewable and sustainable sources for production value added to products related to functional biomolecules, food supplements, biofuels, and applications for wastewater treatment. This renewed interest among pharmacologists and algae biotechnologists alike was due to isolation

DOI: 10.1201/9781003322115-2

and identification of novel molecules with potential uses as high-value nutraceuticals or potential drug analogues from microalgae. For the past decade, microalgae are increasingly explored on par with fungi, bacteria, and higher plants for novel drug molecules with high potential for commercialization. The reason for this interest was due to the presence of new metabolites with unique chemical structures that cannot be found in higher plants or can be produced using genetically engineered bacterial platforms. Due to their unique evolutionary origin, many microalgae species, especially diatoms, contain molecules that are found not only in plant systems but also in marine invertebrates, fungus, and bacteria. Despite their enormous potential to date, very few microalgae-based technologies have developed into successful commercial systems. The reason for this is non-availability of microalgae strains that can reach high productivity, low-cost production systems, availability of technologies to produce multiple products from biomass, and improvement in biomass and biomolecule production under varied geographical locations. So, in this chapter we address the significance of diatom algae for sustainable biorefinery for high-value compounds and the main challenges involved in their production. While diatoms are widely present in terms of diversity and abundance in nature, few species are currently used for biotechnological applications. This is most likely due to difficulties in their cultivation. Our research work on diatoms for the past decade has made us realize their enormous potential as precursors of many important, valuable metabolites. So, through this chapter we want to address the key drawbacks that are hindering exploitation of diatoms for product development like strain selection, media for maximum growth and bio-molecule production, and cost-effective mass culturing techniques.

2.2 BIOTECHNOLOGICAL SIGNIFICANCE OF DIATOMS

Diatoms are a key eukaryotic microalgae group and are one of the most common and dominant type of phytoplankton in aquatic ecosystems (Armbrust, 2009). They contribute up to 45% of the total oceanic primary production and 25% of global oxygen output in oceanic systems. Marine diatoms are responsible for up to 20% of the global CO_2 fixation (Wilhelm et al, 2006). The fact that diatoms have overtaken other phytoplankton in nature proves their superiority over other phytoplankton.

Diatoms produce a myriad of isoprenoid compounds owing to their unique evolutionary, physiological, and metabolic potential. This makes them exciting prospects for an extensive array of industrial and commercial applications. Isoprenoids are the class of ubiquitous secondary metabolites encompassing many biotechnologically pertinent molecules finding established applications in pharmaceutical, cosmetics, food, and fuel industries. The isoprenoids in diatoms include the carotenoids like fucoxanthin, sterols like ergosterol, highly branched isoprenoids (HBIs), and squalene, which are antioxidants, anti-obesity, anti-inflammatory, cytostatic and anti-HIV agents, immune stimulators, and biofuel feed stocks. The low toxicity levels of these biomolecules derived from diatoms make them safe pharmaceutical ingredients, facilitating their application in human health and therapeutics. Despite these attractive, unique attributes diatoms are least explored in terms of microalgae commercial systems compared to other algae classes. The principal reasons for the underutilization are due to non-availability of diverse strains that can reach high cell densities in mass culture systems and efficient strategies to enhance and sustain high biomolecule content with the advent of novel strategies for strain selection and improvement and the availability of genomic and transcriptomic data.

Diatoms can biosynthesize omega-3 long chain polyunsaturated fatty acids [(n − 3) LC-PUFA], like arachidonic acid (ARA; 20:4(n-6)), eicosapentaenoic acid (EPA; 20:5(n − 3)), and docosahexaenoic acid (DHA; 22:6(n −3)) (Lebeau et al., 2003). These essential fatty acids are not naturally produced by invertebrates and mammals, so they are incorporated in marine animals through their diet and in humans through consumption of omega-3 rich fish oils, which acquire them by consuming phytoplankton. These fatty acids are recognized as essential for survival and growth during the early stages of life of many marine invertebrates, which constitute the secondary and tertiary part of marine food webs. In humans, omega-3 and omega-6 PUFA, especially DHA and ARA, are essential for infant nutrition for brain and retinal development and are important fatty acids in human breast milk. They form an integral part of commercially available infant food formulas.

Mansour et al. (2005) reported that diatoms contained higher EPA content compared to chlorophytes. Bumbak et al. (2011) reported that researchers at martek biosciences (USA) are growing EPA-rich *Nitzschia alba* heterotrophically. Medina et al. (1999) showed that *P. tricornutum* is a potential source of EPA due to its faster growth rate and presence of low concentration of other LC-PUFA like DHA and ARA, which might interfere during extraction and purification. Medina et al. (1999) showed that the two main techniques for concentrating and purifying diatom-derived PUFA were urea fractionation and HPLC. Belarbi et al. (2000) achieved over 70% of the EPA content with *P. tricornutum,* and the purity of diatom-derived EPA is higher compared to fish oil. These articles show the potential of diatom biomass for EPA production. The present commercial EPA source is fish oil, and due to high fluctuation of price and sustainability issues, the cost of production is gradually increasing. The current market price of bulk EPA is around $650/kg (Belarbi et al., 2000). Fish oil is known to contain high heavy metal contents, especially mercury, and as fish bio-magnify endocrine-disrupting chemicals like pesticides, there is an increasing risk that oils derived from fish might carry over their contamination. Thus, an alternative economical and sustainable source of EPA is essential. Japan is one of the major markets for EPA and DHA supplements, and annual EPA demand in Japan is 125 tones. With increased clinical studies on the biological importance of PUFA for human health, the demand for supplements is expected to increase.

Fucoxanthin (FX) is a unique carotenoid, and it is one of the most abundant carotenoids present in marine brown seaweed. FX has a unique chemical structure that contains both hydroxyl gatherings and epoxide bond alongside an allenic bond and a carbonyl group conjugated with 5,6-momoepoxide in the polyene chain combined with an acetyl group, which is responsible for its bioactivity. Fucoxanthin is an orange-colored pigment, and it acts as an auxiliary pigment in combination with chlorophylls a and c in brown seaweeds (Phaeophyceae) and diatoms (Bacillariophyta). Diatoms have a characteristic golden-brown color due to a presence of equal amounts of chl a and FX in their light-harvesting complex. Fucoxanthin is known to exert its numerous health benefits in humans (Peng et al., 2011). Traditionally seaweeds are consumed as part of the diet in South-East Asia and some European countries, but the contents of FX are very low in seaweed (Kim et al., 2012). Therefore, searching for alternative sources of fucoxanthin is necessary. Diatom algae contain 1.5%–2.5% of FX as dry cell weight (DCW), which is significantly higher than any other natural source. In diatoms, FX functions as light-harvesting pigment not like other bioactive pigments like astaxanthin, which are storage compounds induced during stress conditions. This aspect is advantageous as there is no need to induce stress for PUFA for FX production and exponentially growing diatoms contain high FX content. This ensures better process control in commercial production systems as induction of stress often results in reduced biomass productivity. Currently, the diatom *Phaeodactylum tricornutum* is the only major natural commercial fucoxanthin source in microalgae due to its high fucoxanthin content (49–59 mg g^{-1} of dry cell weight). But the biomass productivity is considerably low, leading to low FX productivity (4–7 mg L^{-1} d^{-1}), which makes it economically unviable for large-scale commercial production.

Diatom extracts are known to exhibit an antibacterial effect against fish and shellfish pathogens. The diatom *Skeletonema costatum* extracts inhibited Vibrio sp., which have a harmful effect on the survival of economically important Penaeus monodon (Naviner et al., 1999). The diatom *Halea ostrearia* is a common diatom species found in oyster farms, and it is known to produce marennine, a blue-colored pigment, in response to nutrient stress, which has shown antitumor activity against human lung cancer cells and anti-HIV effects (Berge et al., 1997). Along with their antibacterial and antiviral properties, especially in aquacultured species, live diatoms contribute significantly to bivalve feeds, which has a market value of around US$100 million (Barnum, 1998).

Present aquaculture practices, especially in developing countries, generate wastewater that is high in nutrients originating from unutilized feeds and aquacultured species excreta. Aquaculture wastewater contains a high amount of organic and inorganic macronutrients like nitrogen and phosphorus, and a variety of trace elements (K, Ca, Mg, Fe, Cu, and Mn). Fortunately, almost all of these organic and inorganic nutrients are essential for the growth and metabolism of microalgae. This aspect makes aquaculture wastewater an ideal environment for growing microalgae, especially diatoms (Li et al., 2011). Similarly, from studies using diluted primary treated sewage effluent, the diatom *P. tricornutum* can successfully remove up to 80% of ammonium and 70% of orthophosphate (Rupert et al., 1995). So, diatoms are an ideal type of algae that can be used for phycoremediation of wastewater in aquaculture ponds because they not only help in nutrient removal but also act as food for cultured organisms, acting as a double-edged sword in maintaining water quality and natural food production. Thompson et al. (2002) found that in wastewater diatom-dominated biofilms are efficient in phosphorus removal compared to conventional methods based on aerobic bacteria. Lefebvre et al. (1996) reported a novel way to recycle excess nutrients from fish farms by growing diatoms and using the biomass as feed for bivalves. This innovative approach is efficient in recycling of excess nutrients and generation of a wealth from wastewater as a circular bioeconomy option. These studies provide evidence for the enormous potential of diatoms in nutrient removal from pollutes from aquatic habitats.

Our understanding of physiology-combined advances in omics and genetic engineering has given new horizons to algal biotechnology in the past three decades. Commercial-scale microalgae cultivation systems are rapidly evolving with identification and characterization of strains that can grow in different trophic modes and can reach very high biomass density. But still only five genera of microalgae are currently explored, namely *Chlorella* sp. and *Spirulina* sp. for protein-based food supplements, *Dunaliella* sp. for β-carotene, *Haematococcus* sp. for astaxanthin, *Schizochytrium* sp. for DHA, and *Phaeodactylum* sp. for fucoxanthin production. Out of this list, the diatom *Phaeodactylum* sp. is a model species with availability of whole genome and metabolic maps for almost all the key processes involved in storage biomolecule biosynthesis. But other species of diatoms that are known to attain higher biomass and biomolecule productivity is not given enough research impotence, leading to gross underrepresentation of diatoms in commercial production systems. The possible reasons can be attributed to the inclination of phycologists toward chlorophytes due to their similarity with terrestrial plants, the difficulty of obtaining samples, and the dynamic nature of oceanic systems where diatoms dominate other microalgae groups.

2.3 PHYCOPROSPECTING AND STRAIN ENGINEERING

Diatoms are highly diverse, and it was reported that an estimate of around 10,000 species are present, of which only a handful of species are explored in commercial scales. Diatoms have substantial evolutionarily-based differences in cellular organization and metabolic processes relative to chlorophytes. This is due to their unique evolution by a secondary endosymbiotic event. Diatoms are the most dominant form of algae in marine environments. Diatoms have tremendous ecological success, with typically higher productivity than other algal classes. Diatoms can accumulate lipid equivalently or to a greater extent than other algal classes, and they contain a high amount of highly potent biomolecules like fucoxanthin. Diatom algae have been grown for decades in aquaculture hatcheries as live feeds in production systems ranging from open ponds to photobioreactors. One of the first available algae whole genome sequences was of diatom *Thalassiosira sp.*, after which many model species like *Pheodactylum tricornutum* were increasingly explored for omics and genetic manipulation approaches for understanding numerous metabolic processes involved in growth and biomolecule production. Despite these attractive attributes, diatoms are seriously underrepresented both in basic and applied research, especially regarding commercial-scale culturing and its related value-added product generation. One of the key drawbacks that we need to address is the availability of strains other than the model organisms that can reach higher biomass density under varied growth conditions and can biosynthesize commercially valuable products in higher concentration.

In the year 1998, the NREL-funded aquatic species program was one of the first research projects that screened microalgae for estimating their potential for biofuel production. Of a pool of 3,000 strains, 50 stains were shortlisted based on critical criteria like growth, lipid productivity, amiability to growth in mass culture systems, and ability to tolerate and adapt to diverse physicochemical conditions. Interestingly, of these shortlisted strains 60% are diatoms. Despite reports like these and enormous research articles on attributes of diatoms, their underrepresentation in scale-up and pilot scale studies is quite startling.

Leabeu et al. (2003) reported the potential of diatoms for biotechnologically relevant product development and production and their ability to attain high productivity even in open pond systems with minimum contamination and tolerance to varying nutrient dynamics for purposes other than biofuel production. A comparative study on growth rate between different groups of microalgae has shown that, diatoms can attain biomass productivity from 5 to 25 g m^{-2} d^{-1}, which is equal to values obtained from other high productive species like Eustigmatophyte *Nannochloropsis salina,* the cyanobacterium *Spirulina platensis* (Thomas et al., 1978). *Cyclotella cryptica,* a centric marine diatom, can reach mean biomass productivity of 20 g m^{-2} d^{-1}, which is the highest reported productivity by a photosynthetic diatom (Huesemann et al., 2009). From India, Suman et al. (2012) reported the potential of native marine diatom isolate *Cylindrotheca clostridium* for EPA production and optimized conditions for maximum growth using a nano-silica-based nutrient mixture. These results indicate the enormous potential of diatoms in generating high biomass in controlled conditions. Fucoxanthin is a photosynthetic axillary pigment in diatoms and a major carotenoid in seaweeds, which is a high-value biomolecule with proven anticancer, antioxidant, antidiabetic, anti-obesity, and anti-inflammatory properties (Woo et al., 2010). *Odontella aurita,* a marine centric diatom, is studied for fucoxanthin production, and the results have shown that low light and high nitrogen concentration facilitates higher fucoxanthin productivity (Xia et al., 2013). This species was also studied for its growth in bubble column reactors, which resulted in biomass and fucoxanthin content of 6.36 g L^{-1} and 18.47 mg g^{-1}, respectively. Kim et al. (2012) reported the potential of model diatom *Pheodactylum tricornutum* as a commercial source of fucoxanthin in which they have identified fucoxanthin as a major carotenoid in this diatom. Fucoxanthin content in the extracts produced by the different methods using varied diatom strains ranged between 15.42 and 16.51 mg g^{-1}. These findings suggest that diatom fucoxanthin volumetric content is about ten times higher than present commercial source marine brown seaweed, macroalgae.

2.4 FACTORS INFLUENCING HIGH CELL DENSITY CULTURING

The growth and cell density of diatoms are influenced by a variety of variables, both biotic and abiotic, which interact in a variety of ways. The evolutionary adaption of species; the development phase; the accessibility of silica, phosphorous, and nitrogen; the silica concentration; temperature; and light intensity are some of the parameters that play a role in this process (Yang et al., 2020).

2.4.1 Trophic Mode

Diatoms are autotrophic creatures that produce their own food via photosynthesis and are thus reliant on light for their energy needs. However, the development of the cells may be stifled by a light shield or photo inhibited regardless of the amount of light present, which would result in a poor biomass yield. To cultivate in either a heterotrophic or mixotrophic manner, several diatoms make use of organic carbon sources like glucose, glycerol, fructose, and acetate (Yang et al., 2020). In most cases, the amount of biomass and bioactive chemicals produced is increased when organic substrates are used. For instance, as compared to the results of autotrophic culture in *Nitzschia laevis,* the yield and productivity of eicosapentaenoic acid (EPA) grown on mixed organic substrates were 9.53 and 8.07 times higher, respectively, than those obtained from autotrophic cultivation (Patel et al., 2019). Microalgae known as diatoms are autotrophic organisms that, given the right circumstances, can extract nutrition from organic carbon sources. Generally, these chemicals are used in the process of energy storage in addition to meeting the metabolic requirements for cell development (Shishlyannikov et al., 2014). Some species of microalgae, for instance, can assimilate up to 85% of the glucose that is obtainable in the form of stored polysaccharides. Even though other species of diatoms are unable to assimilate glucose, they possess the enzymatic activity required for their metabolic activities (Yang et al., 2020). A few diatoms can use glycerol that is present in the medium, which leads to a rise in their growth rate and triggers various biochemical and morphological modifications inside the cell, most notably in the shape of chloroplasts (Shishlyannikov et al., 2014).

Employing glycerol and urea as their corresponding carbon and nitrogen sources, *P. tricornutum* was found to exhibit a behavior that was comparable to the one described earlier (Wang et al., 2020). The effectiveness of biomass and eicosapentaenoic acid synthesis reached respective levels of 1.52 g $L^{-1}d^{-1}$ and 42.13 mg $L^{-1}d^{-1}$. The addition of glycerol, as shown by transcriptomic analysis, led to a 2.27-fold increase in the expression of stearoyldesaturase, an enzyme that both contributes to the process of eicosapentaenoic acid production in *P. tricornutum* and encourages the storage of eicosapentaenoic acid (Villanova et al., 2017). Experimental studies with the addition of 13C-glycerol demonstrate that more than 50% of the carbon in *P. tricornutum* was sourced from glycerol, whereas the residual carbon was sourced from the CO_2 fixation process (Huang et al., 2015). This may be one factor that contributes to a decrease in volumentric content of photosynthetic pigments under mixotrophic conditions (Liu et al., 2009). It is worthwhile to know that the different strains of diatoms have distinct pathways for the conversion of carbohydrate (Zheng et al., 2013). In *Synedra acus*, the addition of glucose led to a substantial rise in biomass, as well as a buildup of chrysolaminarin and eicosapentaenoic acid (Shishlyannikov et al., 2014). Furthermore, under heterotrophic circumstances, *N. laevis* can make utilization of glucose to boost both the rate of cell development and the amount of eicosapentaenoic acid and fucoxanthin that it contains (Guo et al., 2016; Lu et al., 2018). *C. cryptica, C. meneghiniana*, and *Navicula incerta* are all able to successfully thrive on glucose in a heterotrophic environment (Wen and Chen, 2000). However, the introduction of glucose drastically decreased the levels of expression of genes encoding PSY, PDS, ZDS, CRISO, LCYB, and ZEP in *P. tricornutum*, which led to a decrease in the amount of fucoxanthin produced by the organism.

On the other hand, glucose has the potential to increase both the growth rate and the generation of biomass in the *P. tricornutum* UTEX 640 strain. Furthermore, organic sources that are produced from low-cost substrates may be utilized in the production of value-added goods. Mixotrophy was an option for growth for *P. tricornutum* SAG 1090-6, as shown by their ability to acquire 19.69 mg L^{-1} per day of EPA production while employing spruce hydrolysate (Patel et al., 2019). Palm oil mill effluent that contains organics may stimulate the proliferation of *P. tricornutum* CCMP2558 cells and may also contribute to the synthesis of fucoxanthin (Nur et al., 2019). It is important to highlight that organic substrate typically contributes to the biomass formation more than they do to the composition of the biologically active substances, hence encouraging the creation of value-added molecules.

2.4.2 Low Light and Blue/Green Light Induction

When microalgae are grown in an autotrophic environment, the light source is the most important and growth-limiting condition (Yang et al., 2020). When it comes to the formation of value-added chemicals, low light intensity is often ideal; nevertheless, the optimum light intensity and exposure to light will vary depending on the species of diatom. Because of the suppression of glycolysis-related genes and the overexpression of BGS1 in *P. tricornutum* (Heydarizadeh et al., 2019), chrysolaminarin biosynthesis was enhanced when the plant was exposed to low sunlight. In *P. tricornutum*, the level of eicosapentaenoic acid and fucoxanthin rose by 5.72% and four times, correspondingly, when the plant was exposed to dim light (Conceição et al., 2020). In *Entomoneis paludosa, Nitzschia alexandrina*, and *Staurosira sp.*, similar effects were found; particularly, carotenoids and eicosapentaenoic acid content dropped as the light intensity rose (Conceição et al., 2020). Since the zeaxanthin cycle and the diadinoxanthin cycle are two xanthophyll cycles in diatoms that take part in non-photochemical quenching under high light intensity, the expression level of violaxanthin de-epoxidase was raised to boost the conversion to zeaxanthin and diatoxanthin, resulting in a drop in fucoxanthin content. When light concentrations were high, the transcript levels of most chlorophyll-a and fucoxanthin biosynthesis-related genes were down-regulated (Nymark et al., 2009). As a result, the accumulation of chrysolaminarin, eicosapentaenoic acid, and fucoxanthin in diatoms benefited from low light circumstances.

Additionally, the various qualities of light play a vital part in the process of cell proliferation and the formation of biological substances (Yang et al., 2020). When it came to diatoms, red light was more beneficial for the creation of biomass, whereas blue light was more beneficial for the deposition of lipids and pigments (Jungandreas et al., 2014). When compared to white and red light, eicosapentaenoic acid production under blue light was able to achieve its greatest level, which was 33,4 mg/L per day, with 0.95% dry weight fucoxanthin content (Marella and Tiwari, 2020). When compared to red light, blue light exhibited greater photoprotective capability, energy transfer efficiencies, and photosystem II (PSII) repair capabilities in *P. tricornutum* (Valle et al., 2014). Furthermore, blue LED lighting may save roughly 50% of the power input than to red-blue LED lighting and 75% of the power input related to white fluorescent lighting.

The unique shape of fucoxanthin–chlorophyll–protein gives rise to the hypothesis that it can collect blue and green light even when it is in deep water, which lends substantial support to the concept of using blue-green light in the synthetic bioreactor. In response to blue light, the levels of expression of PSY, PDS, ZDS, and ZEP increased, which resulted in an increase in fucoxanthin content (Valle et al., 2014). Consequently, a blend of white and blue light with a ratio of 1:1 might also boost fucoxanthin production in *N. laevis*, which would result in the greatest degree of productivity yet documented. In a two-phase culture, the amount of eicosapentaenoic acid that was produced rose greatly when green light was used as the inducing agent; however, the amount of carotenoid that was produced reduced after three days of induction (Jung et al., 2019). Earlier research showed that the maximum thresholds of lycopene β-cyclase, violaxanthin deepoxidase, zeaxanthin epoxidase-1, and zeaxanthin epoxidase-2 transcripts were accomplished in *Phaeodactylum tricornutum* following induction with green light for 0.5 hours (Valle et al., 2014). However, the beneficial effects of green light were no longer present after the plant was exposed to green light for more than 24 hours. Cell development and the deposition of high-value chemicals in diatoms are both dependent on effective light intensity and quality, but these two factors do not occur simultaneously. Therefore, to increase the production of biomass and high-value chemicals in diatoms, a two-step method that improves the light intensity and light quality may be adopted (Jung et al., 2019). This technique involves increasing the intensity of the light while also improving the quality of the light.

2.4.3 Nutrient Requirements

Be it a small-scale laboratory batch culture or an industrial-scale fed-batch or continuous culture, certain factors like light, temperature, and pH, besides maintaining an aseptic environment, essentially regulate the growth and productivity of the culture. Moreover, the nutrient composition of any culture media is one of the most important factors that governs the productivity of the culture. In the case of microalgal cultures, besides water, many macronutrients and micronutrients play a crucial role in the cell growth and biosynthesis of products. The macronutrients that form an essential part of the nutrient media for culturing diatoms include carbon, nitrogen, phosphorus, sulphur, magnesium, and silica, where silica forms an indispensable constituent of the nutrient media for the growth of diatoms with siliceous frustules. Micronutrients like iron, manganese, vanadium, boron, molybdenum, copper, nickel, cobalt, and selenium are also important constituents of any culture media.

Carbon sources such as glucose and acetate have been found to be highly efficient in increasing the microalgal biomass (Chen and Chen, 2006). Carbon not only acts as a source of energy for heterotrophic growth but also helps the carbon skeleton of the microalgae. Glucose is a well-known energy-rich compound but acetate, as reported by *Cyclotella cryptica* (CCMP 333) showed the highest concentration of fucoxanthin in autotrophic as well as heterotrophic conditions when glucose was used as a carbon source (Guo et al., 2016). This augmentation can be explained by the Acetyl-CoA metabolism for fatty acid synthesis. While conversion of glucose to acetyl-CoA requires several metabolic steps, acetate converts directly into acetyl-CoA by acetyl-CoA synthetase. Other carbon sources that can be utilized include ethanol, glycerol, fructose, etc. depending on the species strains to be cultured.

Besides carbon, phosphorus is another essential macroelement that needs to be supplied in the culture media for the proper growth of the biomass. Phosphorus forms the part of many important biomolecules, especially the backbone of nucleic acids apart from phospholipids of the biological membranes, enzymes, and energy transfer systems (Pahl, 2010). Phosphates are also essential in the synthesis of astaxanthin and PUFAs in algal systems. Microalgae are capable of utilizing an organic form of phosphates, but when conducting experiments in vitro, phosphorus is generally provided in the form of inorganic orthophosphate. A study on *P. tricornutum* has confirmed the role of phosphate in the regulation of specific genes responsible for transcriptional upregulation of sodium-dependent phosphate cotransporters in phosphorus-depleted media (Sharma et al., 2019).

Another most important macronutrient on the list is nitrogen. Nitrogen is an essential component of the amino acids, proteins, and the nitrogenous bases of DNA and RNA, which are important signal-transducing biomolecules. Thus, an adequate supply of nitrogen is an important requisite while culturing microalgal individuals. Nitrogen sources can be of two types, viz., simple nitrogen sources and complex nitrogen sources. While the former includes nitrate and urea, the latter comprises tryptone, yeast extract, and corn steep liquor (Chen and Chen, 2006). Besides urea, cells that die during culturing also serve as an important source of NH_4^+ (Orefice et al., 2019). Complex nitrogen sources are preferred over simple sources as they directly supply important organic nitrogen in the form of amino acids and vitamins.

While culturing diatoms, silica is one of the components that must be present in the culture medium. It is not only an indispensable element of the culture medium but also acts as a limiting agent. It is generally supplied either in the form of sodium metasilicate pentahydrate ($Na_2SiO_3.5H_2O$) or sodium metasilicate nonahydrate ($Na_2SiO_3.9H_2O$). These forms in solution convert into silicic acid ($Si(OH)_4$), and upon being metabolized, they get converted into and deposited as the hydrated form of silica ($SiO_2.nH_2O$) to form the characteristic diatom frustule (Pahl, 2010).

Above all, the concentration of dissolved oxygen in the medium is essential for both photoautotrophic as well as heterotrophic cultivation systems. Regulation of dissolved oxygen concentration must be taken care of as high oxygen concentration inhibits algal growth and productivity in photoautotrophic cultures, while low concentration inhibits the growth and productivity of heterotrophic systems. Moreover, increased dissolved oxygen concentration has been seen to increase fatty acid desaturase activity, thereby leading to elevated unsaturated fatty acid production (Buono et al., 2016, Pahl, 2010).

The most commonly used culture media for culturing coastal, marine microalgae is F/2 media, which is half the concentration of the original "f-medium" by Guillard and Ryther (1962). It is a seawater-enriched nutrient solution especially designed for the culturing of diatoms. Other trusted culture media for diatoms include Walne medium and Conway medium. The ratio of C, N, and P has been defined as 106:16:1 by Alfred Redfield for the optimum growth of microalgal biomass. Since diatoms require silica for the development of their frustules, Si has been added, and this composition ratio of C:N:P: Si has been modified to 106:16:1:16 (Sarthou et al., 2005).

A comparative account of standard medium for culturing diatoms that represent their nutrient composition in terms of different major and minor nutrients, vitamins for cultivating diatoms and attaining maximum biomass are listed in Table 2.1.

TABLE 2.1
Culture Medium Protocols and Analysis of Diatom Growth

Nutrient Composition	F/2-Si Medium (mg/L)	Walne Medium (mg/L)	WC Medium (mg/L)	PM Medium (mg/L)
Macronutrients	$Na_2SiO_3.9H_2O$ – 30 $NaH_2PO_4.H_2O$ – 5 $NaNO_3$ – 75	$CuSO_4.5H_20$ – 2000 $NaH_2PO_4.H_2O$ – 20000 $(NH_4)6Mo_7).4H_20$ – 900 $NaNO_3$ – 10000	$CaCl_2$ – 36.76 K_2HPO_4 – 08.71 $MgSO_4.7H_2O$ – 36.97 $Na_2SiO_3.9H_2O$ – 28.42 $NaHCO_3$ – 12.60 $NaNO_3$ – 85.01	$CaCl_2$ – 0.37 K_2HPO_4 – 2.90 $MgSO_4.7H_2O$ – 3.70 $Na_2SiO_3.9H_2O$ – 14.21 $NaHCO_3$ – 3.15 $NaNO_3$ – 56.7
Micronutrients	Na_2EDTA – 4.360 $FeCl_3.6H_2O$ – 3.150 $CuSO_4.5H_2O$ – 0.010 $MnCl_2.4H_2O$ – 0.180 $ZnSO_4.7H_2O$ – 0.022 $CoCl_2.6H_2O$ – 0.010 $Na_2MoO_4.2H_2O$ – 0.006	Na_2EDTA – 4.50 $FeCl_3.6H_2O$ – 1.30 $CuSO_4.5H_2O$ – 2.00 $MnCl_2.4H_2O$ – 4.00 $ZnCl_2$ – 2.10 $CoCl_2.6H_2O$ – 2.00 H_3BO_3 – 3.36	Na_2EDTA – 4.360 $FeCl_3.6H_2O$ – 3.150 $CuSO_4.5H_2O$ – 0.010 $MnCl_2.4H_2O$ – 0.180 $ZnSO_4.7H_2O$ – 0.022 $CoCl_2.6H_2O$ – 0.010 $Na_2MoO_4.2H_2O$ – 0.006 H_3BO_3 _ 1.000	Na_2EDTA – 2.180 $FeCl_3.6H_2O$ – 1.580 $CuSO_4.5H_2O$ – 0.005 $MnCl_2.4H_2O$ – 0.090 $ZnSO_4.7H_2O$ – 0.011 $CoCl_2.6H_2O$ – 0.005 $Na_2MoO_4.2H_2O$ – 0.003 H_3BO_3 – 0.500
Diatom species & their maximum biomass productivity ($mgL^{-1}day^{-1}$)	Chaetoceros sp. =125 ± 3 & 750 ± 3 Skeletonema sp. = 185.3 ± 2.63 & 562.5 ± 3.96 Thalassiosira sp. = 312.5 ± 2.51 & 433.5 ± 1.80	*Skeletonema* sp. = 160.83×10^4	*Sellaphora minima* = 60 *Nitzschia palea* = 150	Skeletonema costatum = 3510 ± 0.17 Chaetoceros calcitrans = 2030 ± 0.06
Reference	Saxena et al., 2020	Endar et al., 2012	Gérin et al., 2020	Bastos et al., 2022

2.4.4 Photobioreactor Selection and Design

To date, open systems are the most widespread methods across the globe to commercially cultivate algae because of their low investment cost. However, PBRs with high capital investment succeed in producing high-quality results as compared to open systems (tanks, circular, and raceway ponds) due to certain drawbacks like moisture loss due to evaporation, exposure to environmental variations in temperature, higher dependence on light optima, humidity, low diffusion of CO_2, and risk of contamination make them a strong competitor to produce high-value compounds in a short span of time. Therefore, to achieve maximum biomass productivity, the selection of the best operating reactor design is the most important strategy because diatoms can grow in all three trophic modes (Villanova and Spetea, 2021). The feasibility study of various diatom strains in different types of PBRs to generate high-value compounds depends on several factors like the mode of culture (continuous/semi-continuous), light condition, nutrient, and silica composition and operational cost with respect to biomass productivity (Fiddy et al., 2022).

The mode of cultivation is paramount to attain industrial-scale biomolecules by using carbon as an energy source, both under light and dark conditions. The design of the reactor should maintain a large area for enhanced cell growth due to high photosynthetic efficiency and easier harvest of biomolecules as large cells sink at a faster rate (Marella et al., 2021). Also, the reactor design should be able to penetrate light to a large depth, to avoid self-shading by designing large columns, and to withstand temperature variations. According to Eilertsen et al. (2022), the growth rates and biovolume of diatom cells were doubled when the size of the reactor increased during a study period from 2015 to 2021. They have scaled the reactor size from 14,000 L capacity to 3,00,000 L, and reported cell size was sustained at 300 m^3 density and retained the lipid content without breaking down. However, optimum light conditions and continuous monitoring of nutrient flow are the prerequisites of a closed set-up to achieve high saturation levels with a high amount of reproducible chemical content (fatty acids and amino acids) in the cells.

Another reactor design using low-density polyethylene (LDPE) wrapped around PBR to test the permeability of gases and production of biofuel from *Nitzschia palea* cultures (Gordon et al., 2019). The LDPE-sealed PBR experiment was tested with different samples like control (without any seal, open PBR), normal polyethylene, and high-density polyethylene (HDPE). The findings suggested that bubble wrap of PBR with LDPE supports diatom growth, and the highest cell count with maximum lipid content of 35 $\mu g\ mL^{-1}$ was observed on the 40th day (Khan et al., 2022). This depicts that sealing the PBR rim with LDPE not only yields maximum high-value content but also retains water and nutrients successfully for many days. Therefore, cutting the operating cost in small-scale production under natural stressed conditions.

The helical tubular reactor design also led to an enhancement of the cell growth rate, by improving metabolic byproduct production during a two-stage process. For example, the high-density production of EPA in *Skeletonema costatum* is observed at the low temperature in a tight cylindro-conical system (0.3 m^3) exposed to continuous flow of sterile air. Similarly, *Cyclotella cryptica* was cultivated in a 15-l photobioreactor for silica production (Csögör et al., 1999) in carefully controlled conditions, to obtain diatoms uniform in shape and size, as required for industrial applications, especially in nanotechnology (Xiang et al., 2017). Similarly, light diffusion was found to be higher in tubular PBR, EPA production rise to 67% in *Phaeodactylum tricornutum* (Acien Fernandez et al., 2000), and Marennine production doubled in *Haslea ostrearia* when the tube diameter was kept at 0.1 m (Molina Grima et al., 2000).

In airlift PBRs with internal illumination systems are being considered very effective in metabolite production. The light wavelength can be adjusted according to the species' growth requirement. For example, *Amphora capitellata* yield maximum chlorophyll-a concentration per unit biomass (2.62 ± 0.16 mg L^{-1}) under red light, which was only 14% higher than under blue light with a specific growth rate of 0.317 day^{-1}, doubling every 2–3 days. Also, fatty acid lipid content rises with decreasing growth rate for *A. capitellata,* suggesting white light has the potential to support the lipid content (Deniz et al., 2019). This technique is easy to maintain due to its smaller surface-to-volume ratio. It is cost-effective to grow diatoms using low-priced organic carbon sources like glycerol and hydrolysates and needs low oxygen intake to generate EPA and DHA.

The PBRs design can be altered as per the availability of the diatom strain, and their requirements for growth like nutrient flow, light availability, organic carbon supplementation, photooxidative stress levels, and temperature fluctuations, during the exponential phase and lag phase can be easily regulated and monitored (Villanova et al., 2017). Despite numerous advantages, further research is needed on looking for novel diatom strains that can be exploited commercially to yield high-value compounds while using multiple cheap carbon sources like lignocellulose materials and food waste hydrolysates coupled with design optimization of existing PBRs to switch between trophic modes (Wang and Seibert, 2017).

2.5 SCIENTOMETRIC ANALYSIS FOR DIATOM BIOMOLECULES AND GROWTH

A total of 13,180 documents with "diatom/s" or "Bacillariophyta" keywords in the title, abstract, or manuscript were sourced from the online database of the "Web of Science" from the year 2010 to the present date (15 May 2022). The articles were categorized as research papers, review papers, correspondence, short communication and conference abstract, errata, and editorials in a total of 80 journals (Ho and Kahn, 2014). The total documents were further shortlisted based on the multidisciplinary but prolific keywords, such as "diatoms," "bacillariophyta," "pennate," "centric," "marine," "freshwater," "biodiversity," "environmental sciences," "forensic science," "ecology," and "toxicology". Also, the impact factors (IF) and cite score were tracked as per the Scopus tools to identify topics and trends (https://academic-accelerator.com/Citescore/Tracking). The top 20 most active journals on diatom research are represented in Figure 2.1, and their impact factor and cite score are mentioned in Table 2.2.

The scientometric analysis has emphasized diatom studies taken in the last decade and their application in the environment. Biological studies will generate qualitative and quantitative data to support the current status and prospects for further studies on a global scale.

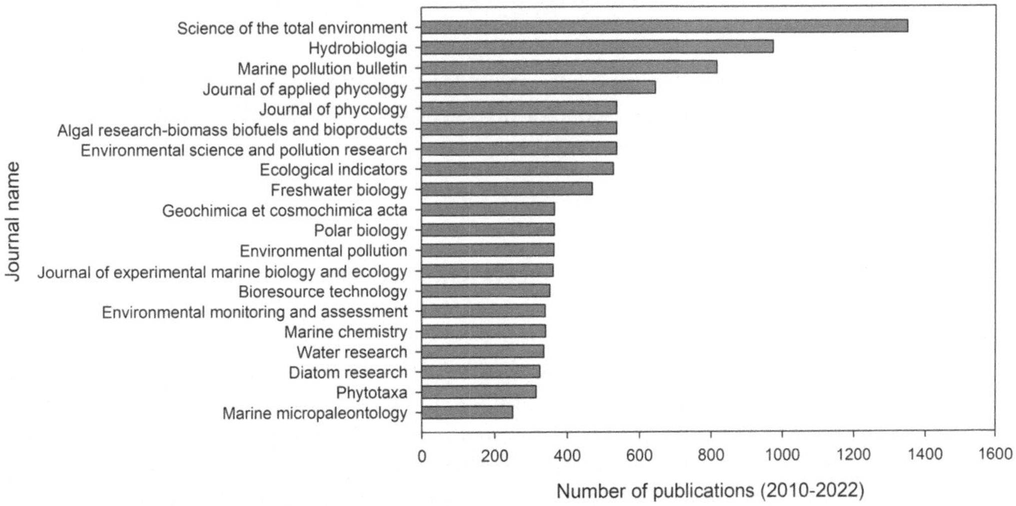

FIGURE 2.1 List of top 20 journals on diatom research from 2010–2022.

TABLE 2.2
Bibliographic Record of Top 20 Journals with Number of Publications, Impact Factor, and Cite Score for the Year 2020

Sr. No.	Journal Name	Number of Publications (2010–2022)	Sum of Impact Factor (2020)	Cite Score (2020)
1	Science of the total environment	1352	7.963	10.5
2	Hydrobiologia	974	2.694	5.1
3	Marine pollution bulletin	817	5.553	7.9
4	Journal of applied phycology	645	3.215	
5	Journal of phycology	537	2.923	4.3
6	Algal research-biomass biofuels and bioproducts	537	4.401	6.9
7	Environmental science and pollution research	537	4.223	5.5
8	Ecological indicators	528	4.958	7.5
9	Freshwater biology	470	3.809	
10	Geochimica et cosmochimica acta	365	5.01	8.6
11	Polar biology	364	2.31	3.6
12	Environmental pollution	363	8.071	10.8
13	Journal of experimental marine biology and ecology	361	2.171	4.8
14	Bioresource technology	352	9.642	14.8
15	Environmental monitoring and assessment	340	2.513	3.6
16	Marine chemistry	339	3.807	5.6
17	Water research	335	11.236	15.6
18	Diatom research	324	1.815	2.5
19	Phytotaxa	314	1.171	1.7
20	Marine micropaleontology	250	2.415	3.6

Moreover, the bibliographic findings help in understanding the knowledge gaps in diatom research and raise awareness about the scope of divergent trends in interdisciplinary areas (Zhang et al., 2019).

2.6 CONCLUSION

The diatom-based functional biomolecule market is a high growth market but in contrast to other microalgae their exploitation is confined to limited biotechnological applications. At present, diatom biomass is mainly used for semi-industrial production of EPA and fucoxanthin along with their use as live feed in aquaculture. Their use in the field of nanobiotechnology for nano-scale silica production as new age drug carriers is not fully explored. The primary reasons for this under exploitation is due to non-availability of strains that can reach high productivity, especially in industrial-scale production systems. So, there is an urgent need to address these issues by finding novel strains through extensive phycoprospecting from varied environments, including extremophilic species that can attain high growth rate and biomolecule content in both mixotrophic and heterotrophic modes. This will address critical drawbacks for their commercialization like photoinhibition and sustainability issues in presently available PBR systems. In the post genomic era, with availability of whole genomes of numerous diatom strains, metabolic processes involved in nutrient acquisition and biomolecule synthesis are increasingly elucidated. This is critical in metabolic engineering of diatoms that can revolutionize their use in nutrition, pharmaceutical, biofuels, and aquaculture. Ongoing advances in genetic manipulation, combined with novel cultivation techniques, will further promote diatoms as a sustainable platform for new age high-value biomolecules.

REFERENCES

Acien Fernandez, F. G. F., Garcia Camacho, J. A. Sanchez Perez, J. M. *et al.* 2000. Modelling of eicosapentaenoic acid (EPA) production from Phaeodactylum tricornutum cultures in tubular photobioreactors. Effects of dilution rate, tube diameter and solar irradiance. *Biotechnology and Bioengineering*, 68:173–183. 10.1002/(SICI)1097-0290(20000420)68

Armbrust, E. V. 2009. The life of diatom in the world,s oceans. *Nature*, 459:185–192.

Barnum, S. R. 1998. *Biotechnology: An introduction*. Wadsworth, Belmont, California.

Bastos, C. R. V., Maia, I. B., Pereira, H. *et al.* 2022. Optimisation of biomass production and nutritional value of two marine diatoms (Bacillariophyceae), *Skeletonema costatum* and *Chaetoceros calcitrans*. *Biology*, 11:594. 10.3390/biology11040594

Belarbi, E-h., Grima, M. E., Chisti, Y. 2000. A Process for high yield and scaleable recovery of high purity eicosapentaenoic acid esters from microalgae and fish oil. *Enzyme Micobial Technology*, 26:516–529.

Berge, J. P., Bourgougnon, N., Carbonnelle, D., Le Bert, V., Tomasoni, C., Durand, P., Roussakis, C. 1997. Antiproliferative effects of an organic extract from the marine diatom *Skeletonema costatum* (Grev.) Cleve. against a non-small-cell bronchopulmonary carcinoma line (NSCLC-N6). *Anticancer Research*, 17:2115–2120.

Bumbak, Fabian, Cook, Stella, Zachleder, Vilém, Hauser, Silas, & Kovar, Karin (2011). Best practices in heterotrophic high-cell-density microalgal processes: achievements, potential and possible limitations. Applied Microbiology and Biotechnology, 91, 31–4610.1007/s00253-011-3311-6.

Buono, S., Colucci, A., Angelini, A. *et al.* 2016. Productivity and biochemical composition of *Tetradesmus obliquus* and *Phaeodactylum tricornutum*: Effects of different cultivation approaches. *Journal of Applied Phycology*, 28:3179–3192. 10.1007/s10811-016-0876-6

Chen, G. Q., Chen, F. 2006. Growing phototrophic cells without light. *Biotechnology Letters*, 28:607–616. 10.1007/s10529-006-0025-4

Conceição, D., Lopes, R. G., Derner, R. B. *et al.* 2020. The effect of light intensity on the production and accumulation of pigments and fatty acids in *Phaeodactylum tricornutum*. *Journal of Applied Phycology*, 32:1017–1025. 10.1007/s10811-019-02001-6

Csögör, Z., Melgar, D., Schmidt, K. *et al.* 1999. Production and particle characterization of the frustules of *Cyclotella cryptica* in comparison with siliceous earth. *Progress in Industrial Microbiology*, 35:71–75. 10.1016/S0079-6352(99)80100-3

Deniz, I., Demirel, Z., Imamoglu, E. *et al.* 2019. Enhanced Microalgal lipid production in internally illuminated airlift photobioreactor. *Marine Technology Society Journal* 53(2):38–45. 10.4031/MTSJ.53.2.4

Eilertsen, H. C., Eriksen, G. K., Bergum, J.-S. *et al.* 2022. Mass cultivation of microalgae: I. Experiences with vertical column airlift photobioreactors, diatoms and CO_2 sequestration. *Applied Science*, 12:3082. 10.3390/app12063082

Endar, V., Sarjito, S., Hutabarat, J. *et al.* 2012. Effect of using Guillard and Walne technical culture media on growth and fatty acid profiles of microalgae *Skeletonema Sp.* in mass culture. *Journal of Coastal Zone Management*, 16 (1):50–56.

Fiddy S. P., Martin, F., Jean-Sébastien Deschênes. *et al.* 2022. Semi-continuous system for benthic diatom cultivation and marennine production. *Algal Research*, 62:102633. 10.1016/j.algal.2022.102633

Gérin, S., Delhez, T., Corato, A. et al. 2020. A novel culture medium for freshwater diatoms promotes efficient photoautotrophic batch production of biomass, fucoxanthin, and eicosapentaenoic acid. *Journal of Applied Phycology*, 32:1581–1596. 10.1007/s10811-020-02097-1

Gordon, R., Merz, C. R., Gurke, S., Schoefs, B. 2019. Bubble farming: Scalable microcosms for diatom biofuel and the next Green Revolution [Chapter 22]. In: Seckbach J., Gordon R. (eds.) *Diatoms: Fundamentals & Applications [DIFA, Volume 1 in the series: Diatoms: Biology & Applications, series editors: Richard Gordon & Joseph Seckbach]*. Wiley-Scrivener, Beverly, pp. 583–654.

Guillard, RRL, & Ryther, JH (1962). Studies of marine planktonic diatoms: I. Cyclotella nana Hustedt, and Detonula confervacea (Cleve) Gran. Can J Microbiol, 8, 229–39. 10.1139/m62-029

Guo, B., Liu, B., Yang, B. *et al.* 2016. Screening of diatom strains and characterization of *Cyclotella cryptica* as a potential Fucoxanthin producer. *Marine Drugs*, 14:125. 10.3390/md14070125

Heydarizadeh, P., Veidl, B., Huang, B. *et al.* 2019. Carbon orientation in the diatom *Phaeodactylum tricornutum*: The effects of carbon limitation and photon flux density. *Frontiers in Plant Science*, 10:471. 10.3389/FPLS.2019.00471/BIBTEX

Ho, Y. S., Kahn, M. 2014. A bibliometric study of highly cited reviews in the Science Citation Index expande. *Journal of the Association for Information Science and Technology*, 65(2):372–385. 10.1002/asi.22974

Huang, A., Liu, L., Yang, C. *et al.* 2015. *Phaeodactylum tricornutum* photorespiration takes part in glycerol metabolism and is important for nitrogen-limited response. *Biotechnology for Biofuels*, 8:73. 10.1186/s13068-015-0256-5

Huesemann, M. H., Hausmann, T. S., Bartha, R. et al. 2009. Biomass productivities in wild type and pigment mutant of *Cyclotella sp.* (Diatom). *Applied Biochemistry and Biotechnology*, 157(3):507–526. 10.1007/s12010-008-8298-9

Jung, J. H., Sirisuk, P., Ra, C. H. *et al.* 2019. Effects of green LED light and three stresses on biomass and lipid accumulation with two-phase culture of microalgae. *Process Biochemistry*, 77:93–99. 10.1016/J.PROCBIO.2018.11.014

Jungandreas, A., Costa, B. S., Jakob, T. *et al.* 2014. The acclimation of *Phaeodactylum tricornutum* to blue and red light does not influence the photosynthetic light reaction but strongly disturbs the carbon allocation pattern. *PLoS One*, *9*:e99727. 10.1371/JOURNAL.PONE.0099727

Khan, M. J., Gordon, R., Varjani, S., Vinayak, V. 2022. Employing newly developed plastic bubble wrap technique for biofuel production from diatoms cultivated in discarded plastic waste. *Science of The Total Environment*, 823:153667. 10.1016/j.scitotenv.2022.153667

Kim, J. C., Heo, J. M., Mullan, B. P., Pluske, J. R. 2012. Performance and intestinal responses to dehulling and inclusion level of Australian sweet lupins (Lupinus angustifolius L.) in diets for weaner pigs. *Animal Feed Science and Technology*, 172(3–4): 201–209. 10.1016/j.anifeedsci.2012.01.002

Lebeau, T., Robert, J. M. 2003. Diatom cultivation and biotechnologically relevant products. Part I: Cultivation at various scales. *Applied Microbiology and Biotechnology*, 60:612–623. 10.1007/s00253-002-1176-4

Lefebvre, S., Hussenot, J., Brossard, N. 1996. Water treatment of land-based fish farm effluents by outdoor culture of marine diatoms. *Journal of Applied Phycology*, 8:193–200. 10.1007/BF02184971

Li, Y., Chen, F. Y., Chen, P. et al. 2011. Characterization of a microalga *Chlorella sp.* well adapted to highly concentrated municipal wastewater for nutrient removal and biodiesel production. *Bioresource Technology*, 102(8):5138–5144.

Liu, X.-J., Duan, S.-S., Li, A.-F. *et al.* 2009. Effects of glycerol on the fluorescence spectra and chloroplast ultrastructure of *Phaeodactylum tricornutum* (Bacillariophyta). *Journal of Integrative Plant Biology*, 51:272–278. 10.1111/j.1744-7909.2008.00767.x

Lu, X., Sun, H., Zhao, W. et al. 2018. A hetero-photoautotrophic two-stage cultivation process for production of fucoxanthin by the marine Diatom *Nitzschia laevis*. *Marine Drugs*, 16:219. 10.3390/md16070219

Mansour, M. P., Frampton, D. M., Nichols, P. D. et al. 2005. Lipid and fatty acid yield of nine stationary-phase microalgae: Applications and unusual C24–C28 polyunsaturated fatty acids. *Journal of Applied Phycology*, 17:287–300. 10.1007/s10811-005-6625-x

Marella, T. K., Bhattacharjya, R., Tiwari, A. 2021. Impact of organic carbon acquisition on growth and functional biomolecule production in diatoms. *Microbial Cell Factories*, 135. 10.1186/s12934-021-01627-x

Marella, T. K., Tiwari, A. 2020. Marine diatom *Thalassiosira weissflogii* based biorefinery for co-production of eicosapentaenoic acid and fucoxanthin. *Bioresource Technology*, 307:123245. 10.1016/J.BIORTECH.2020.123245

Molina Grima, E., Acien Fernandez, F. G., Garcia, C. F. *et al.* 2000. Scale up of tubular photobioreactors. *Journal of Applied Phycology*, 12:355–368.

Naviner, M., Bergé, J-P., Durand, P., Le Bris, H. 1999. Antibacterial activity of the marine diatom *Skeletonema costatum* against aquacultural pathogens. *Aquaculture*, 174(1–2):15–24. 10.1016/S0044-8486(98)00513-4

Nur, M. M. A., Muizelaar, W., Boelen, P., Buma, A. G. J. 2019. Environmental and nutrient conditions influence fucoxanthin productivity of the marine diatom *Phaeodactylum tricornutum* grown on palm oil mill effluent. *Journal of Applied Phycology*, 31:111–122. 10.1007/S10811-018-1563-6/FIGURES/6

Nymark, M., Valle, K. C., Brembu. *et al.* 2009. An integrated analysis of molecular acclimation to high light in the marine diatom *Phaeodactylum tricornutum*. *PLoS One*, 4:e7743. 10.1371/JOURNAL.PONE.0007743

Orefice, I., Musella, M., Smerilli, A. *et al.* 2019. Role of nutrient concentrations and water movement on diatom's productivity in culture. *Scientific Reports*, 9(1):1–10. 10.1038/s41598-018-37611-6

Pahl, S. L., Lewis, D. M., Chen, F., King, K. D. 2010. Heterotrophic growth and nutritional aspects of the diatom *Cyclotella cryptica* (Bacillariophyceae): Effect of some environmental factors. *Journal of Bioscience and Bioengineering*, 109:235–239.

Patel, A., Matsakas, L., Hrůzová, K. *et al.* 2019. Biosynthesis of nutraceutical fatty acids by the oleaginous marine microalgae *Phaeodactylum tricornutum* utilizing hydrolysates from organosolv-pretreated birch and spruce biomass. *Marine Drugs*, 17:119. 10.3390/md17020119

Peng, J., Yuan, J.-P., Wu, C.-F., Wang, J.-H. 2011. Fucoxanthin, a marine carotenoid present in brown seaweeds and diatoms: Metabolism and bioactivities relevant to human health. *Marine Drugs*, 9:1806–1828. 10.3390/md9101806

Robles Medina, A., Esteban Cerdán, L., Giménez Giménez, A. et al. 1999. Lipase-catalyzed esterification of glycerol and polyunsaturated fatty acids from fish and microalgae oils. *Marine Bioprocess Engineering*, Proceedings of an International Symposium Organized Under Auspices of The Working Party on Applied Biocatalysis of the Eurpean Federation of Biotechnology and The European Society for Marine Biotechnology, 379–391. 10.1016/s0079-6352(99)80131-3

Rupert J. C., Valerie J. S., Paul J. M. 1995. Wastewater nutrient removal by marine microalgae cultured under ambient conditions in mini ponds. *Water Science and Technology*, 31(12):151–160. 10.1016/0273-1223(95)00502-E.WPOO

Sarthou, G., Timmermans, K. R., Blain, S. *et al.* 2005 Growth physiology and fate of diatoms in the ocean: A review. *Journal of Sea Research*, 53:25–42.

Saxena A., Prakash K., Phogat S., Singh P. K., Tiwari A. 2020. Inductively coupled plasma nanosilica-based growth method for enhanced biomass production in marine diatom algae. *Bioresource Technology*, 314:123747. 10.1016/j.biortech.2020.123747

Sharma, A. K., Mühlroth, A., Jouhet, J. *et al.* 2019. The Myb-like transcription factor Phosphorus Starvation Response (PtPSR) controls conditional P acquisition and remodeling in marine microalgae. *New Phytologist*, 225:2380–2395. 10.1111/nph.16248

Shishlyannikov, S. M., Klimenkov, I. V., Bedoshvili, Y. D. *et al.* 2014. Effect of mixotrophic growth on the ultrastructure and fatty acid composition of the diatom *Synedra acus* from Lake Baikal. *Journal of Biological Research-Thessaloniki*, 21:15. 10.1186/2241-5793-21-15

Suman, K., Kiran, T., Devi, U. K., Sarma, N. S. 2012. Culture medium optimization and lipid profiling of *Cylindrotheca*, a lipid-and polyunsaturated fatty acid-rich pennate diatom and potential source of eicosapentaenoic acid. *Botanica Marina*, 55:289–299. 10.1016/j.biortech.2019.02.031

Thomas, W. H., Dodson, A. N., Reid, F. M. H. 1978. Diatom productivity compared to other algae in natural marine phytoplankton assemblages2. *Journal of Phycology*, 14(3), 250–253. 10.1111/j.1529-8817.1978.tb00294.x

Thompson, R. K., McKinnon, J. J., Mustafa, A. F. et al. 2002. Chemical composition, ruminal kinetic parameters, and nutrient digestibility of ammonia treated oat hulls. *Canadian Journal of Animal Science*, 82(1):103–109. 10.4141/A01-036

Valle, K. C., Nymark, M., Aamot, I. *et al.* 2014. System Responses to Equal Doses of Photosynthetically Usable Radiation of Blue, Green, and Red Light in the Marine Diatom *Phaeodactylum tricornutum*. *PLoS One*, 9:e114211. 10.1371/JOURNAL.PONE.0114211

Villanova, V., Fortunato, A. E., Singh, D. *et al.* 2017. Investigating mixotrophic metabolism in the model diatom *Phaeodactylum tricornutum*. *Philosophical Transactions of the Royal Society B: Biological Sciences*, 372. 10.1098/RSTB.2016.0404

Villanova, V., Spetea, C. 2021. Mixotrophy in diatoms: Molecular mechanism and industrial potential. *Physiologia Plantarum*, 173(2):603– 611. 10.1111/ppl.13471

Wang, J. K., Seibert, M. 2017. Prospects for commercial production of diatoms. *Biotechnology and Biofuels*, 10:16. 10.1186/s13068-017-0699-y

Wang, X., Balamurugan, S., Liu, S.-F. *et al.* 2020. Enhanced polyunsaturated fatty acid production using food wastes and biofuels byproducts by an evolved strain of *Phaeodactylum tricornutum*. *Bioresource Technology*, 296: 122351. 10.1016/j.biortech.2019.122351

Wen, Z. Y., Chen, F. 2000. Production potential of eicosapentaenoic acid by the diatom *Nitzschia laevis*. *Biotechnology Letters*, 22:727–733. 10.1023/A:1005666219163

Wilhelm, C., Büchel, C., Fisahn, J. *et al.* 2006. The regulation of carbon and nutrient assimilation in diatoms is significantly different from green algae. *Protist*. 157(2):91–124. 10.1016/j.protis.2006.02.003

Woo, M.-N., Jeon, S.-M., Kim, H.-J. et al. 2010. Fucoxanthin supplementation improves plasma and hepatic lipid metabolism and blood glucose concentration in high-fat fed C57BL/6N mice. *Chemico-Biological Interactions*, 186(3), 316–322. 10.1016/j.cbi.2010.05.006

Xia, S., Wang, K., Wan, L., Li, A., Hu *et al.* 2013. Production, characterization, and antioxidant activity of fucoxanthin from the marine diatom *Odontella aurita. Marine drugs*, *11*(7), 2667–2681. 10.3390/md11072667

Xiang, X., Ozkan, A., Chiriboga, O. *et al.* 2017. Techno-economic analysis of glucosamine and lipid production from marine diatom *Cyclotella sp. Bioresource Technology*, 244: 1480–1488. 10.1016/j.biortech.2017.05.079

Yang, R., Wei, D., Xie, J. 2020. Diatoms as cell factories for high-value products: Chrysolaminarin, eicosapentaenoic acid, and fucoxanthin. *Critical Reviews in Biotechnology*, 40, 993–1009. 10.1080/07388551.2020.1805402

Zhang, Y., Tao, J., Wang, J. *et al.* 2019. Trends in diatom research since 1991 based on topic modeling. *Microorganisms*, 7(8), 213. 10.3390/microorganisms708021

Zheng, Y., Quinn, A. H., Sriram, G. 2013. Experimental evidence and isotopomer analysis of mixotrophic glucose metabolism in the marine diatom *Phaeodactylum tricornutum. Microbial Cell Factories*, 12, 109. 10.1186/1475-2859-12-109

3 Ecological Impact of Diatoms

Mitigation to Indicators

Noor Haziqah Kamaludin
Occupational Health and Safety Risk Management (OHSeRM) Research Initiative Group & Center of Environmental Health & Safety, Faculty of Health Sciences, Universiti Teknologi MARA, Puncak Alam, Selangor, Malaysia

Nur Azalina Suzianti Feisal and Mohd Azwan Ahmad
Department of Diagnostic and Allied Health Science, Faculty of Health and Life Sciences, Management and Science University, Selangor, Malaysia

Tengku Nilam Baizura Ibrahim
Occupational Health and Safety Risk Management (OHSeRM) Research Initiative Group & Department of Environmental Health, Faculty of Health Sciences, Universiti Teknologi MARA, Cawangan Pulau Pinang, Kampus Bertam, Kepala Batas, Pulau Pinang, Malaysia

CONTENTS

DOI: 10.1201/9781003322115-3

3.1 INTRODUCTION

Extensive analyses on various aspects of diatom biology, ecology, and evolution have emerged throughout the years. Diatom-based ecological indicators have great attraction among researchers, environmental professionals, and the community. They have already been employed to identify variations in the environment, but scientific maturation in indicator advancement has mostly developed over the last 20 years. Indicators are currently employed, mainly to analyze the quality of the environment, as early detection of ecological problems, and as key indicators for variations in ecological resources. In recent years, a great diversity of diatom-based categorization approaches has been developed globally. These current technologies, along with the oldest, evolved from the necessity to analyze the water quality of rivers, but understanding of potential divergences caused by them was insufficient.

Diatoms are photosynthesizing microalgae with a diameter between 2 and 500 microns that are characterized by a siliceous skeleton (frustules) and can be found in a wide range of aquatic habitats, including freshwater and saltwater, soils, and almost everywhere there is moisture. There are diatoms that can exist in such great abundance and be sufficiently preserved as sediments almost exclusively made of the frustules of these diatom (diatomite) forms; these deposits are economically valuable because diatoms are used in filter media, paints, toothpaste, and more. Diatom species have become an immensely useful instrument for monitoring previous and current environmental conditions. This can contribute to water quality and climate change studies.

3.1.1 Classification of Diatoms

Diatom cell walls are made up of two valves that fit together like pill capsules. The valves are also crucial in their identification and classification. Simonsen (1979) found that diatoms can be classified into two orders which are Centrales and Pennales. The Centrales, also known as Biddulphiales, contain valve striae that are oriented radially in respect to a point, annulus, or central areola. Meanwhile, Pennales and Bacillariales have valve striae that are aligned along a line and seem bilaterally symmetrical. According to Round et al. (1990) and Hoek et al. (1995), diatoms can be divided into three classes: centric diatoms (Coscinodiscophyceae), pennate diatoms without a raphe (Fragilariophyceae), and pennate diatoms with a raphe (Bacillariophyceae).

3.2 LIFE CYCLE OF DIATOMS

Diatoms often split and reproduce by a process known as vegetative division, which includes the division of a unicellular organism into two new cells. The new cell is generated inside the original cell throughout the reproduction cycle. Because it originates within the mother cell, which has a solid cell wall that does not extend, the new cell is smaller in size. During this phase, the new cell adopts a valve from the original frustule as its epitheca before constructing its hypotheca in around 15 minutes. This procedure can be performed several times each day (1 to 8 times). This, however, is extensively relative to the amount of dissolved silica.

As the average size of the diatom frustule in a population decreases, restoration of the frustule size becomes important. At this phase, auxospores are generated. After nutrition levels have been reduced, the first auxospore cell generates a new frustule of optimum size, which then develops after intensive vegetative reproduction. The cycle continues as the nutrient level rises and a thicker-walled resting cyst develops in the statospore stage. As a result, seasonal upwelling plays a key role in the life cycle of many diatoms, not only as a source of food but also as a means to transport statospores and their vegetative products to the photic zone.

3.3 DIATOMS AS AN INDICATOR

Excessive nutrients lead to a degraded lake ecosystem, stressing the importance of comprehensive nutrient removal systems that allow us to conserve or remediate bodies of water. To monitor nutrient variations, biological components such as macroinvertebrates, macrophytes, diatoms, and zooplankton have been exploited (Lougheed et al., 2007). Because of their sensitivity and time-dependent reactivity, diatoms are one of the most extensively investigated species for water quality assessment globally (Stevenson, 2014). Ecological indicators are often used to analyze water quality, hydrology, and the general health of an aquatic environment depending on the biodiversity of these microalgae. Toxin levels, physicochemical properties, and the nature of the water supply are evaluated using indicator species (Nixon, 2009).

Diatoms are important in the generation of biomass and the absorption of greenhouse gases in water. Diatoms are responsible for approximately 20% of global photosynthetic CO_2 fixation, which is comparable to the photosynthetic rate of all rainforests altogether, and around 40% of yearly oceanic biomass production (Falkowski et al., 1998). Diatoms are incredibly resilient and can survive in almost all photic regions from the equator to the arctic, in which they are widely explored for their use as biomarkers of changes in physiochemical circumstances resulting in a rapid responsiveness to even minor changes (Datta et al., 2019). Hence, diatoms exhibit a wide range of adaptability in a variety of environments that might be beneficial for their application in biotechnological processes regardless of stressful conditions. Diatoms are responsive to fluctuations in their aquatic ecosystem and function as excellent indicators of water quality. This is due to their high reproductive rate, which allows for a substantial growth in population of a specific species under optimal circumstances while other species diminish or vanish.

3.3.1 Eutrophication

Eutrophication is the supplementation of water bodies with inorganic plant nutrients (Khan and Mohammad, 2014). Lake eutrophication is the process when nutrient sources, typically phosphorus (P) and nitrogen (N), exceed the levels that would happen as a result of any system disturbance, resulting in enhanced lake productivity. Human (anthropogenic eutrophication) and environmental (natural eutrophication) disruptions are both causes of eutrophication. Natural eutrophication occurrences are particularly unusual but can happen as a consequence of drastic incidents such as forest fire and tree die-off (Hall and Smol, 2010). Extreme weather events like droughts, for example, may concentrate lakewater nutrients or enhance the input of nutrient-rich groundwater (Mulholland and Webster, 2010). Anthropogenic nutrient inputs from household and industrial sewage disposal, agricultural activities, and surface runoff have been the most prominent causes of water quality challenges (Datta et al., 2019; Gottschalk et al., 2011). On a worldwide basis, eutrophication is the most frequent cause of lake pollution, and it has numerous detrimental impacts on water ecosystems (Harper, 1992). Eutrophication affects biochemical processes and biodiversity in order to improve total produced output.

At various levels of the food chain, significant shifts occur, and whole populations might transform or die. Changes in the phosphorus (P) and nitrogen (N) ratio frequently result in a shift in primary production from diatoms and other smaller edible algae rather than toward bigger cyanobacteria that are stronger competitors for nitrogen (N) and more resistant to grazing (Hall and Smol, 2010). Many shallow reservoirs may destroy their submerged freshwater macrophyte populations when the earth heats up with increasing phytoplankton turbidity, affecting higher trophic levels. The decomposition of plant and algal biomass restricts the oxygen supply in deep waters, resulting in a reduction in suitable aquatic habitats and, in severe cases, humongous fish kills. The critical issue is that eutrophication may aggravate water ecosystems to become instable, and biodiversity can be destroyed (Hall and Smol, 2010).

3.3.2 Nitrification

Nitrification is the process in the nitrogen cycle that integrates the oxidation of ammonia, which is generated by the decomposition of organic materials to release fixed nitrogen in the form of nitrogen gas. It is conducted by a variety of microorganisms, including ammonia-oxidizing bacteria and archaea, and nitrite-oxidizing bacteria. These microorganisms are predominantly autotrophic aerobes. It is conducted by a variety of microorganisms, including ammonia-oxidizing bacteria and archaea, and nitrite-oxidizing bacteria. These microorganisms are all autotrophic aerobes. Their unique metabolism gives them a special niche, but it also causes delayed and inadequate growth. Anammox organisms are another group of microorganisms that undergoes direct anaerobic oxidation of ammonia to nitrogen gas. Traditional nitrification generally occurs in soils, sediments, and aquatic ecosystems.

Nitrification has an important role in agriculture, since it determines the availability of fertilizers containing nitrogen, and in wastewater treatment, where it contributes to removing excess nitrogen. In low-oxygen waters and sediments, nitrification functions as a significant oxygen sink attributed to denitrification. Nitrifying bacteria are required in the synthesis of the greenhouse gas nitrous oxide in both terrestrial and aquatic ecosystems, primarily through a reductive process recognized as nitrifier denitrification. Environmental factors like salinity, temperature, oxygen, and pH influence nitrification levels in the ecological ecosystem.

3.3.3 Acidification

Our aquatic ecology system is negatively affected by acidification, in particular fish and benthic macroinvertebrates, where climate change activity can alter and induce the concentration of CO_2 and nutrient availability (Andren and Jarlman, 2008; Li et al., 2012). To detect detrimental effects, environmental monitoring sampling and programs are used to monitor fish and macroinvertebrates in the water column (Bach et al., 2019). Diatoms are used as environmental indicators in a variety of ways to determine how acidification is occurring in lakes, rivers, and streams. The trend in planktic diatoms and primary production has increased since pre-industrial times, where it is consistent with the expected climate change impact caused by anthropogenic activities around the world, ranging from catchment-specific pollutants to the global impact.

Benzo(a)pyrene diatoms decreased the seawater acidification through light tolerance in unialgal laboratory cultures of *Pseudo-nitzschia australis (Frenguelli),* one of the toxigenic pennate diatoms capable of releasing the neurotoxin domoic acid (DA), to assess the species' physiological response to ocean acidification (OA), the decline in pH resulting from increased atmospheric CO_2 (pCO_2) (Li et al., 2021; Wingert and Cochlan, 2021). These combined effects of marine pollution and climate-driven ocean acidification are one of the main global problems to marine ecosystems. The diatom assembles the greatest positive potential in terms of global change environmental surveillance where it can balance the ionic composition, acid-base balance and physical properties (Rivera-Rondon and Catalan, 2020). The presence of benthic diatoms can provide an indication of native fish introductions, cattle grazing, and global warming where it supports an adaptive and sustainable management of high-mountain lakes for the sake of nature conservation (Cantonati et al., 2021). Studies using diatoms have clarified a good and positive trend after several environmental sampling monitoring and paleolimnology studies based on the major periods categorized by shifts in diatom composition.

Lakes are increasingly assessed based on their periphyton community. Since respiration tends to be more sensitive to rising temperatures than photosynthesis under the current CO_2 conditions, it is predicted that the efficiency of the biological pump that facilitates carbon sequestration will decline in the future. Epilithic diatoms play a prominent role as bioindicators due to the predominance of shallow exposed waterbodies (Kennedy and Buckley, 2021; Tong et al., 2020). Diatom Acidification Metric (DAM) samples are used to measure the genetic differences between affected and unaffected freshwater acidification that contribute to the issues that afflict vast areas of Europe. These measures are validated at sites with long-term biological and chemical data in a good agreement (Juggins et al., 2016). Overall, diatoms are one of the good suggestions for the ongoing anthropogenic acidification that poses a serious threat to biodiversity, and trophic links in freshwater ecosystems should be considered.

3.3.4 Salinization

Salinization as a result of human activities such as agriculture, resource extraction, and urbanization has now become a major global environmental challenge for the freshwater ecosystem, resulting in considerable biodiversity loss and ecosystem loss (Cunillera-Montcusi et al., 2022; Zhao et al., 2020). The salinization showed the impacts of chloride concentrations in a variety of ways such as low river flows becoming more common from summer to autumn, reducing the dilution of chloride emitted with a steady load of increasing its concentration. It also increased the open water evaporation and reduced rainfall during summer periods, and drought increases the chloride concentration in water and raises the sea level by increasing the seawater intrusion through the tidal closure (Bonte and Zwolsman, 2010). The diatoms served as the basis for assessing the ecological status using the diatom index IO which is used by the state monitoring system.

Previous research found that canonical correspondence analysis (CCA) showed that the axes are 70% of the variance species, and 78% of the variance species have a relationship with the environment where the salinization is associated with the diatom taxa distribution (Bak et al., 2020).

3.4 SENSITIVITY OF DIATOMS TO PHYSICOCHEMICAL CHANGES

The sensitivity of diatoms to physicochemical changes is commonly used to assess the trophic states of the environment media, especially rivers and seawater. The physicochemical trophic state indexes (TSIs) and trophic based on bioindicators, especially periphytic diatoms (TDIs), can be complementary approaches under certain circumstances (Torres-Franco et al., 2019). Diatoms' functional traits reflect the positive changes in the physicochemical changes such as channel width flow, velocity, depth, dissolved oxygen, and water temperature variables to minimize the ecological impact (Wang et al., 2021). In South Korea, the trophic diatom index (TDI) is used to indicate the physicochemical change in the weir section where they were found to be more efficient in term of monitoring and managing the water quality of consecutive weir sections and weir systems (Lee et al., 2021). Nutrient concentration was found to be one of the important diatom trajectories where it can impact the fish predation pressure and climate change in a shorter time (Berthon et al., 2014).

The sensitivity of diatoms to physicochemical changes in the environment is useful based on the quick and precise response of the diatoms for water quality assessment and the substrate-related variability of diatoms in terms of assemblage structure. Diatoms are one of the main groups of organisms used in acid mine drainage (AMD) processes and were the best indicators and had the best sensitivity to evaluate the environmental situation.

3.5 TAXONOMY OF INDICATOR SPECIES FROM DIFFERENT ENVIRONMENTS

Diatoms have long been acknowledged for their effectiveness as environmental indicators. Their enormous abundance and species diversity, which are scattered over all other aquatic ecosystems, contribute to the utilization of their ability. In accumulated sedimentary, their remains are incredibly resilient and well maintained. Most researchers who use the diatom as environmental proxies pay little attention to how and why this species diversity exists in these surroundings. Multiple paleoenvironmental classification systems based on salinity, substrate, and elevation relative to tidal levels have resulted from diatom species preferences for different ecosystems.

Moreover, diatoms' short life cycles enable quick colonization of new species in response to abrupt environmental change. The large number of species and their unique biological distribution provide the groundwork for using them as indicators of nutritional conditions, water transparency, temperature, and pH. Phytoplankton, macroalgae, and diatoms all have ecologically significant growth forms and survival strategies. Periphytic diatom community indicators such as cell density, biovolume, percentage relative abundance, species richness, and Shannon index are commonly utilized for water quality monitoring.

Generally, during the last two decades, the functional conception of diversity has aroused huge interest: it is a component of biodiversity that may be represented as the value and range of taxonomic and individual features that influence diatom functioning. Diatoms were previously classified into three groups: *Coscinodiscophyceae*, *Fragilariophyceae*, and *Bacillariophyceae*, which correlated to three different types of valve configuration. Confidentially, these three structural variants can be classified as "centrics" (*Coscinodiscophyceae*), "araphid pennates" (*Fragilariophyceae*), and "rapid pennates" (*Bacillariophyceae*). The current biogeographic challenge is to reconcile the taxonomic classification of diatoms with a functional classification in order to better anticipate the effects of global changes on ecosystem functions and service environments. Diatoms are known as oxygen factories because they create large amounts of biomass that account for roughly 20% of global net primary oxygen production and have a significant impact on CO_2 levels in the atmosphere. A diatom frustule's form is relatively consistent and an important feature. However, there are some deviations that should be known, as they have in the past led to formation of various varieties, and even new species. These obscenely formed and deformed varieties of diatom behavior are known to be triggered by both physical and chemical surpluses or deficiencies. Although taxonomic diatom records are useful for ecological and empirical studies, precise identification is essential since physically similar species might have ecologically unique ecological attributes.

3.5.1 Diatom Indicators in the Lake Environment

Periphyton is important for biogeochemical cycling and helps to maintain higher trophic levels by providing food webs, which is critically valuable in shallow water bodies. A variety of environmental stresses are threatening the biological integrity of the productive littoral zone, and its state must be evaluated to preserve functioning of ecosystems and lake biodiversity. The sensitivity of lake diatom taxonomy species assemblages to a variety of environmental stressors, including nutrient enrichment, acidification, salinization, metal contamination, pesticides, and hydromorphological change, has been observed. Since the beginning of scientific observation, taxonomic diatom species from the genera *Aulacoseira Thwaites* and *Nitzschia Hassal* have produced seasonal alternating concentrations in the main lake basin. Because diatoms' silicified cell walls aid in their long-term preservation in sediments, they are frequently used to examine water quality and long-term lake reactivity to environmental changes. The benefits of utilizing diatoms as a bioindicator for water quality were that they reflected water quality and the combined effect of many conditions, whereas physico-chemical parameters only revealed water quality when samples were taken and measured.

3.5.2 Diatom Indicators in the Ocean Environment

The huge seasonal cycle in the ocean sea causes large environmental oscillations that alter phytoplankton productivity and ecosystem processes. Deep ocean water with high nutrient levels and a low microbial burden in-situ could be a valuable resource for any phytoplankton culture, including diatoms. In onshore surface sediments, the preponderance of *Chaetoceros spores, T. nitzschioides*, and *F. doliolus* revealed a conserved composition resembling the late boreal autumn and winter trap population. Due to the sheer importance of higher water column diatom dissolution in the oceanic environment and interpretation of fossil diatom assemblages, numerous methods for measuring water column dissolution and diatom preservation have been developed.

3.5.3 Diatom Indicators in the Arctic and Antarctic Environment

The polar regions, including the Arctic and Antarctic, present clear signs of climate change affecting freshwater species, communities, and ecosystems, and are anticipated to continue to change frequently in the future. Antarctic ecosystems are also inhabited by a diverse range of species, many of which are found nowhere else on the planet. Limited water environment in the Arctic and Antarctic stream emerges inside these ice-free regions, nourished by the melting of snow and glacial ice as well as occasional precipitation. Accurate taxonomy is important to use diatoms as environmental indicators to their full potential, with the lowest prediction errors occurring in analyses with high and consistent taxonomic precision. *Cyanobacteria sp.* (also known as blue-green algae) are the most common and numerous freshwater organisms that can be found as individuals, tiny colonies, or cyanobacterial mats in lakes, ponds, meltwater ice streams, and also on moist soils. When compared to temperate locations, the diversity of diatom species in Antarctica is minimal. This was validated using a worldwide taxonomically consistent data set at the genus level, which revealed that lakes in the Antarctic region consistently have lower local and regional diatom richness than homologous water bodies in the northern hemisphere at comparable latitudes.

3.6 WATER QUALITY INDICES OF DIATOMS

Raising awareness of previous and continuing changes in ecosystems caused by human pressure has resulted in the establishment of regulatory norms and legislation to prevent the infinite alteration of the aquatic environment. Since chemical and physical monitoring techniques alone cannot capture all the characteristics important to biota, using indicator organisms to fully monitor the ecological condition of surface waters is a more effective strategy (Figure 3.1).

(Szczepocka et al., 2018). Diatoms will demonstrate a strong correlation as a bio-indicator in evaluating integrated water quality, ecosystem function, and biodiversity, while they are a weak indicator for evaluating ecosystem integrity (Harding and Taylor, 2011).

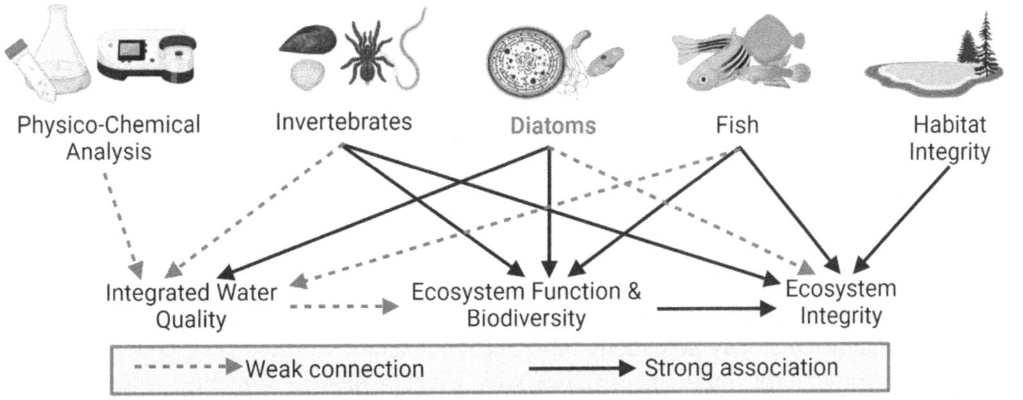

FIGURE 3.1 Aquatic ecosystem monitoring with various types of bio-indicator organisms. The schematic depicts the relationship between bio-indicator organisms with the instruments used to monitor the environment and ecosystem conditions.

Due to the excessive requirement to monitor water quality, standardized sample techniques and evaluation methodologies based on single, simple indicators have been developed. The most widely employed diatom-based indices for assessing water quality are based on traditional taxonomy, genus, and, in many instances, species (Vasselon et al., 2017). These taxonomic diatom indices are based on how diatom assemblages are made up and how many of them there are. The main reason they were made was because each species has different needs for its environment, a different range of where it lives (its "indicator value"), and a different sensitivity to environmental problems (sensitivity value). Most indices are based on the Zelinka and Marvan (1961) formula, which considers species abundances, indicators, and sensitivity levels depending on the system's saprobic condition. Other indices measure trophic, organic, acidification, and metal pollution (Solak, 2011a; Zaghloul et al., 2020; Zelinka, 1961).

The use of diatoms for biological monitoring has a comparatively long history. Indices are mostly developed by means of diatom communities (e.g., diversity, evenness, richness, similarity), so there are often strong links between the different indices. Diatoms are highly suited for water quality studies due to their short generation period and the sensitivity of many species to ecological variables. Several pollution indexes based on diatoms have been made and are being used regularly right now. The new indices are based on a system called saprobic (Szczepocka and Żelazna-Wieczorek, 2018). Diatoms have been utilized as bioindicators in Europe, North America, South America, Asia, and Africa (Reid et al., 1995). Some of these methods are used to figure out what the hydrochemical properties of lakes were in the past, whereas others are used to see how rivers and streams are doing right now.

Since 1902, indices have been used all over Europe to track changes in water quality (Kalyoncu and Akkoz, 2009). However, nothing is known about how these indexes function in Turkey. Kalyoncu and Akkoz (2009) employed the saprobic index to conduct the first study utilizing diatoms to measure water quality changes in Turkey, which was followed by further studies. Furthermore, there are two distinct tests that use different indices. Diatom indices DAIPo (Diatom Assemblage Index for Organic Pollution), SI (Saprobic Index), GI (Generic Index), and TDI (Trophic Diatom Index) have been employed in the Karasu River, Japan and found that GI is very modestly linked with physicochemical parameters. Solak (2011b) used Ominidia Software to analyze SLA (Sládeek Index), DESCY (DESCY Index), LMA (Leclercq and Maquet Index), WAT (Watanabe's Index), CEE (Index of European Economic Community), and IDAP (Artois Picardie Diatom Index) indices and discovered that DESCY appears to be the best for the region. Kalyoncu and Akkoz (2009) applied principal component analysis to look at the 58 diatom taxa used in the saprobic index. They found that SI was a better way to show changes in water quality.

Environmental laws like the Water Framework Directive 2000/60/EC in Europe and the Environmental Monitoring (WFD) and Assessment Program Surface Waters EPA/620/R-94/004F (EMAP-SW) in the United States have been made to protect these aquatic ecosystems (Dalu and Froneman, 2016). For water quality evaluations, EU Member States use previously defined taxonomic indexes that have been intercalibrated across EU Member States (Table 3.1). Due to this, it was possible to apply common indices despite substantial evidence indicating these measures become less useful in locations apart from those where species-environment relationships were first explored. In order to address this biogeographic issue, a number of nations created new indices based on their own variability and the species that lived there.

Regardless of their relevance and widespread usage, taxonomic indexes are associated with several uncertainties that may potentially influence the outcome of the water quality evaluation. This may be particularly important for sites with good to moderate quality, where a wrong classification could waste a lot of time and money or not protect the area well enough. Some of these problems arise from the fact that diatom indexes are made for certain parts of the world but are used a lot in others (Tapolczai et al., 2019). The specific pollution index (IPS) was created in France but is utilized across more than half of the European Union. There are various diatom

TABLE 3.1
List of Diatom Indices for the Assessment of Water Quality by EU Member States

Index Name	Member States Using It	Focus	References
Specific Pollution Index	Most of EU members (Greece, Belgium, Estonia, Luxembourg, Sweden, Finland, Bulgaria, Croatia, Cyprus, Spain, Portugal, Hungary, Slovakia, Poland, Latvia)	Organic pollution	(Juggins et al., 2016)
Metal Pollution Index	Hungary, Italy, Lithuania	Heavy metal pollution	(Fernández et al., 2018)
Duero Diatom Index	Duero basin at NW Spain	Nutrient pollution	(Álvarez-Blanco et al., 2013)
Trophic Diatom Index	United Kingdom, Croatia, Ireland, Poland	Trophic status	(Álvarez-Blanco et al., 2013)
Biological Diatom Index	France, Romania, Spain, Poland	Trophic status	(Coste et al., 2009)
Eutrophication Diatom Index	Italy, Slovakian, Poland	Trophic status	(Kelly, 1998; Kelly & Whitton, 1995)
Generic Diatom Index	Belgium, Poland, NW Spain	Organic pollution and trophic status	(Szczepocka et al., 2018)
Diatom Multimetric Index	Spain	Trophic status	(Carayon et al., 2020; González-Paz et al., 2020; Szczepocka et al., 2018)
Artois-Picardie Diatom Index		General pollution	(Besse-Lototskaya et al., 2011)
Multimetric Diatom Index Trophic Index Saprobic Index	Poland, Slovenia, Germany, Austria, Hungary, Denmark	Trophic status and organic pollution	(Delgado et al., 2012)
Diatom Acidification Metric	United Kingdom, Ireland	Acidification status	(Juggins et al., 2016)

indexes, such as the trophic diatom index (TDI) created in the United Kingdom but utilized in the other three nations, and the biological diatom index (BDI) (developed in France but used in the other three countries) (Beltrami et al., 2012; Dell'uomo andTorrisi, 2011). Widespread use of locally created indices may be difficult since species' reactions to environmental conditions vary by geographic or habitat dispersion. This is shown in ecoregional indexes that employ different ecological profiles for the same species. Taxonomic misidentification happens when similar species have different ecological optimums. Furthermore, since their ecological profiles are not well characterized, the existence of uncommon species or taxa with revised taxonomy is difficult to assess (Kaleli et al., 2020; Tapolczai et al., 2019).

WFD's importance spurred nations outside of Europe to explore adopting comparable laws and evaluation methodologies soon after its establishment (Masouras et al., 2021). Some countries, like Turkey, have made their own biomonitoring indices that are better for their ecoregion (Turkey trophic index) (Solak, 2011a). In North America, biomonitoring is used by the United States under the Clean Water Act, but Canada has taken steps to create a more structured approach for bioassessment methods that are not used very often right now (Lavoie et al., 2014). Argentina in South America has made its own water quality indexes based on diatoms (Pampean diatom index) (Gómez and Licursi, 2001). Before WFD was put in place in Europe, Australia had been using phytobenthos to measure biological quality for a long time, but there was no government coordination (D'Archino and Piazzi, 2021). South Africa has a long history of diatom study and usage in biomonitoring in Africa; nevertheless, more work is required for structured implementation in other African nations. The South African Diatom Index (SADI) is the second biggest index supported by OMNIDIA's current autecological data (Harding and Taylor, 2011). The SADI is built on the same foundation as the majority of European diatom indices, but with a focus on the Specific Pollution Sensitivity Index (SPI) (Table 3.1). The SPI index has the most diverse species basis of any index in use today. This species base has been expanded with the inclusion of indigenous SA diatom species, with scores changed for other global and frequently occurring South African species, boosting the reliability of diatom-based water quality analyses in South Africa (Harding and Taylor, 2011). In Asia, the Asian Pacific Water Summit in 2007 heralded the start of a new era in the assessment of water quality in Asian countries (Table 3.2).

To address the limitations of taxonomic indices, nontaxonomic measurements based on functional features, including cell size, ecological guilds, life-forms, and DNA sequences, have recently evolved, which also includes operational taxonomic units, exact sequence variants, and individual sequence units. Moreover, ways for measuring water quality utilizing benthic diatom assemblages against various environmental parameters, including prediction models and statistical approaches, have been established (e.g., machine learning). Table 3.3 provides all of these intriguing innovative strategies, together with their advantages and drawbacks, which require further scientific examination before being implemented for WFD aims.

TABLE 3.2
Class Limit Boundaries for the SPI

Index Score	Class
$17 \leq SPI \leq 20$	High quality
$13 \leq SPI < 17$	Good quality
$9 \leq SPI < 13$	Moderate quality
$5 \leq SPI < 9$	Poor quality
$SPI < 5$	Bad quality

TABLE 3.3
The Benefits and Drawbacks of Taxonomy-based Diatom Indices Used for Water Quality Evaluation so far, as well as New Interesting Techniques

Indices	Advantages	Disadvantages
Taxonomy-based	• Accurate method • Many references • Many indices	• Time-consuming • Taxonomic misidentification/skilled microscopists • Expensive optical equipment • Access to specialized literature • Lack of rare species database • Different species response in different ecoregions • Need more studies for different ecoregions
Trait-based	• Less time-consuming • Taxonomy free • Less effort • Related environmental stressors	• Need more generalized environmental gradients
Molecular-based	• Less time-consuming • Taxonomy free	• Complementary tool for biomonitoring • Most species not represented in the molecular databases • Cannot be linked to observed species in the field
Predictive models	• Less time-consuming • Taxonomy free	• Need more studies to validate their results
Machine learning techniques	• Less time-consuming • Taxonomy free	• Need more studies

3.6.1 Trait-based Diatom Indices

Trait-based techniques are a potential way to measure ecological status/quality. Traits are "any morphological, physiological, or phenologically observable attribute at the individual level" and are more closely linked to ecosystem functioning than species. Examining them allows a more comprehensive ecological evaluation at the ecosystem level. Environmental pressures (pesticides–herbicides, heavy metals, fertilizers, and organic contaminants) affect benthic diatoms' cell size, ecological guilds, and life-forms (Fragoso et al., 2018; Stenger-Kovács et al., 2018).

Cell size is related to cell biovolume, or the cell's length-to-width ratio. Regardless of the fact that cell size is among the most important functional traits, as it is associated with the update of nutrient and growth rates, there is still no proof that it is affiliated with trophic status or organic pollution based on organic waste; even so, smaller cells are said to enhance at high pesticide concentrations at the expense of larger cells. Ecological guilds are groupings of organisms that share resources and are divided into high, low, motile, and planktic guilds (Stenger-Kovács et al., 2018). Current-induced physical disturbances and nutrient concentrations are linked to ecological guilds. Motile cells favor low-flow and nutrient-rich environments, whereas low profile cells prefer high-flow and nutrient-poor settings. There is a connection between life forms and species that live as solitary cells or colonies, as well as the substrate. Different life forms are associated with trophic status and organic micropollutants, with tube-forming colonial diatoms being more vulnerable to trophic and organic pollution than stalked diatoms (Salmaso et al., 2019; Trábert et al., 2020).

Because these characteristics are relatively easy to measure and characterize, using trait-based indices for river quality assessment should be beneficial because it will require minimal time and energy than taxonomy-based indices (Tapolczai et al., 2017). The same rationale that was used to create the well-known taxonomic-based indices was used to create trait-based indices. Using an environmental gradient of nutrient content and organic micropollutants, the ecological value of each characteristic was assigned, and the resultant distribution was fitted using a model. Trait-based indices, as well as trait-based and taxonomic indices, showed that the former may accurately predict water quality changes (Meyer et al., 2022). Despite the fact that trait-based approaches offer considerable promise for monitoring water quality, they have only been tried in a few rivers, islands, and nations (France). This means that the few trait-based indicators that have been produced so far are highly dependent on the unique environmental variables and gradients present at the research locations. In order to guarantee the robustness of such markers before they are used for regular biomonitoring, further research in various ecoregions and along environmental gradients is required (Alric et al., 2021).

3.6.2 Molecular-based Diatom Indices

High-throughput sequencing (HTS) techniques, particularly DNA metabarcoding, seem to be a useful tool for future monitoring and evaluation processes, giving standardized data that are reliable and high-quantity, but with less time and cost than older taxonomic methods. Similar to a barcode, these methods use gene markers to determine taxa-specific sequences in a cell's DNA. These sequences are saved in a database and ascribed to a species of diatom. As a result, several taxa may be identified using multiple samples at the same time, which speeds up the assessment process (Bailet et al., 2019; Vasselon et al., 2017).

Biomonitoring by HTS of benthic diatoms began with sequence taxonomic identification, enabling previously established indexes to be employed with newly collected sequences (Baird and Hajibabaei, 2012). There is a lack of reference libraries for researching diatoms because of the low number of common taxa found in genomic and morphological databases. Due to the destruction of so much genetic diversity, bioassessment is no longer possible because of the loss of so much genetic diversity. Accordingly, taxonomy-free indices (i.e., DNA indexes that do not take into account taxonomy) are being developed (Apothéloz-Perret-Gentil et al., 2017; Baird and Hajibabaei, 2012).

The DNA data used to generate these indices were classified using a variety of methods, resulting in a wide range of indices with the purpose of avoiding clustering bias. The most common categorization is an operational taxonomic unit (OTU), which groups sequences based on their similarity (Tapolczai et al., 2019). By affecting the number of groups included in the research, the similarity requirements (typically 97–99%) create procedure uncertainty. On the other hand, clustering with high sequence similarity raises the danger of assigning ecological relevance to sequencing mistakes and artefacts. Individual sequence units (ISUs) and exact sequence variations (ESVs) have been proposed as a more defined grouping, allowing for research repeatability and meta-analysis (Kang et al., 2021). There is no categorization or grouping of these unique DNA values with biological relevance. Each sequence group (OTUs, ESVs, ISUs) was positioned along the environmental gradient specified by the supplied dataset to identify its ecological optimum, with the sensitivity and indicator values assigned to the group (Tapolczai et al., 2019).

Taxonomy-free indices yielded the same results as taxonomy-based indices when evaluating water quality, demonstrating their use in routine biomonitoring (Feio et al., 2020). Data from France, Switzerland, and Portugal were the only ones included in the analysis, and the results showed a strong correlation with environmental gradients. In the absence of a comprehensive reference library, the DNA extraction method, the DNA barcode utilized, the bioinformatics treatment, sequence clustering criteria, and the gene copy number per cell bias were the most used HTS techniques. (Masouras et al., 2021). Figure 3.2 provides an understanding of the ecosystem biomonitoring method for determining the water quality using a molecular-based diatom indicator (Baird and Hajibabaei, 2012; Vasselon et al., 2019).

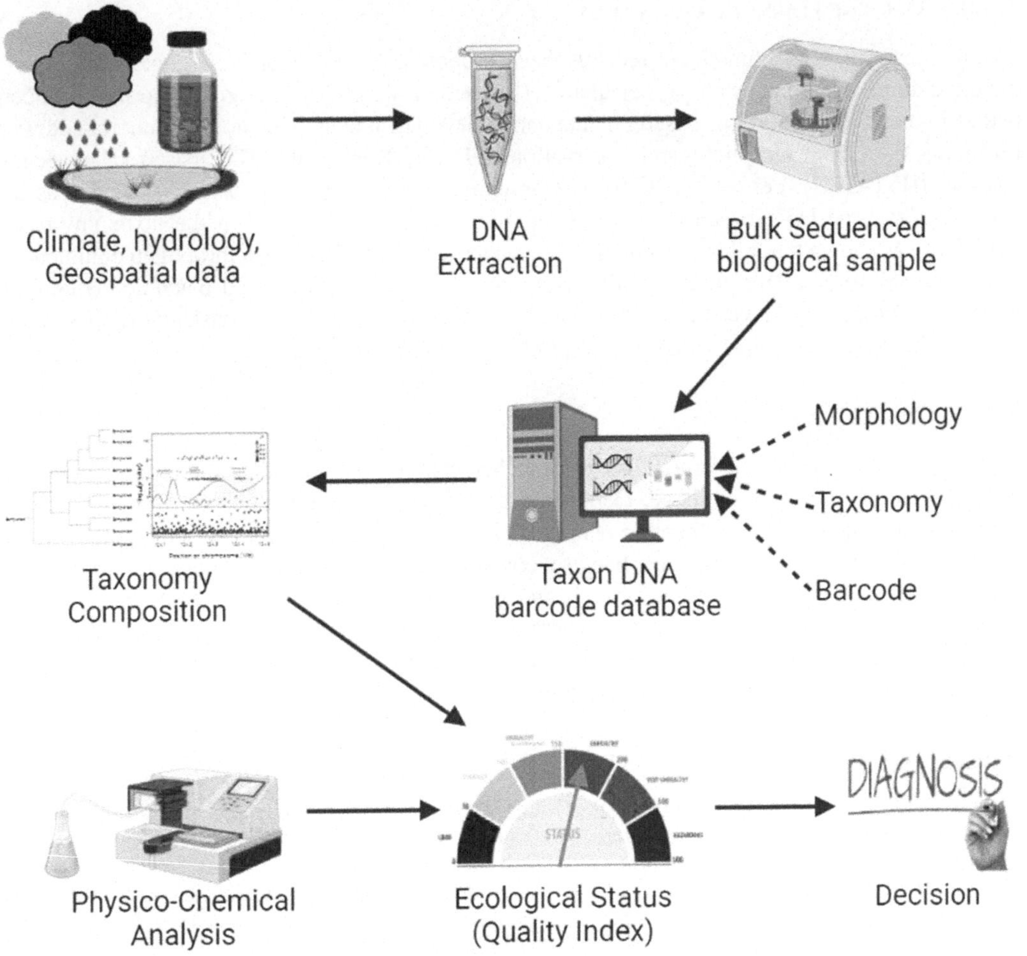

FIGURE 3.2 Ecosystem biomonitoring sequencing using molecular-based diatom indices.

3.6.3 Predictive Models

Predictive models evaluate a site's quality by comparing impacted and reference diatom assemblages under comparable environmental conditions. Because it focuses on the structure and composition of an assemblage rather than a single species, this method is part of a more comprehensive strategy and has been used by Portugal (DIATOMOD) and Northern Spain (NORTIdiat) (Almeida and Feio, 2012).

These diatom-based forecasting algorithms are a way to measure the ecological health of rivers and streams without using taxonomies. Because prediction models compare sites that are affected and sites that are used as a reference in similar situations, it is important to pick reference sites in different types of rivers. It was found that the NORDIdiat model was good at predicting eutrophication and intensive agriculture, whereas the DIATOMOD model was good at predicting hydromorphological change. These results show that it might be possible to use predictive modelling to measure the quality of water using diatoms. However, because of its restricted use and reliance on certain kinds of rivers, it is less beneficial for biomonitoring as a whole than other options available today. In the future, more research with different data sources should be done in other places to confirm these results (i.e., species, traits, or molecular data) (Feio et al., 2012; Masouras et al., 2021).

3.6.4 Machine Learning Techniques

Machine learning algorithms have recently been proposed as an alternative to taxonomy in predicting diatom assemblages and water quality (Busseni et al., 2020). The goal is to make models that can predict diatom groups based on data about the environment. The models were trained with taxonomic data from morphological observations (HYDMORPH) and OTU lists from molecular analyses (HYDGEN) (Feio et al., 2020). The observed-to-predicted richness ratio is used to assess water quality. The OTU-based model was more sensitive to stressors than the taxonomy-based model. It was shown that all these models were accurate and sensitive to anthropogenic influences. This approach to assessing water quality using taxonomy-free datasets and machine learning algorithms is promising; nevertheless, more studies are needed to validate consistent results before they can be used for regular biomonitoring (Almeida and Feio, 2012).

3.7 CONCLUSION

Diatoms are a type of microscopic algae with a characteristic feature called the frustule (siliceous coating). They are diverse in shape. Depending on the habitat, diatoms can be either planktonic, which live in suspended water; benthic, which are associated with substrate; or both planktonic and benthic. In recent decades, there have been several significant efforts all over the world to asses the water quality not only for nutrient, metal, and pesticide parameters but also for biological indicators; however, it is not easy to define where our ecosystem is damaged and polluted. This gives negative results for the ecosystem, such as structure and characteristics, as well as the impact on the health of humans, without further evaluation.

REFERENCES

Almeida, S. F. P., & Feio, M. J. (2012). DIATMOD: Diatom predictive model for quality assessment of Portuguese running waters. *Hydrobiologia*, 695(1), 185–197. 10.1007/s10750-012-1110-4

Alric, B., Dézerald, O., Meyer, A., Billoir, E., Coulaud, R., Larras, F., Mondy, C. P., & Usseglio-Polatera, P. (2021). How diatom-, invertebrate- and fish-based diagnostic tools can support the ecological assessment of rivers in a multi-pressure context: Temporal trends over the past two decades in France. *Science of The Total Environment*, 762, 143915. 10.1016/j.scitotenv.2020.143915

Álvarez-Blanco, I., Blanco, S., Cejudo-Figueiras, C., & Bécares, E. (2013). The Duero diatom index (DDI) for river water quality assessment in NW Spain: Design and validation. *Environmental Monitoring and Assessment*, 185(1), 969–981. 10.1007/s10661-012-2607-z

Andren, C. and Jarlman, A. (2008). Benthic diatoms as indicators of acidity in streams. *Fundamental and Applied Limnology*, 173(3), 237–253. 10.1127/1863-9135/2008/0173-0237

Apothéloz-Perret-Gentil, L., Cordonier, A., Straub, F., Iseli, J., Esling, P., & Pawlowski, J. (2017). Taxonomy-free molecular diatom index for high-throughput eDNA biomonitoring. *Molecular Ecology Resources*, 17(6), 1231–1242. 10.1111/1755-0998.12668

Bach, L. T., Hernandez-Hernandez, N., Taucher, J., Spisla, C., Storna, C., Riebesell, U. & Aristegu, J. (2019). Effects of elevated CO_2 on a natural diatom community in the subtropical NE Atlantic, *Frontiers in Marine Science*, 6, 75. 10.3389/fmars.2019.00075

Bailet, B., Bouchez, A., Franc, A., Frigerio, J.-M., Keck, F., Karjalainen, S.-M., Rimet, F., Schneider, S. C., & Kahlert, M. (2019). Molecular versus morphological data for benthic diatoms biomonitoring in Northern Europe freshwater and consequences for ecological status, Metabarcoding and Metagenomics, 3, e34002. 10.3897/mbmg.3.34002

Baird, D. J., & Hajibabaei, M. (2012). Biomonitoring 2.0: A new paradigm in ecosystem assessment made possible by next-generation DNA sequencing. *Molecular Ecology*, 21(8), 2039–2044. 10.1111/j.1365-294X.2012.05519.x

Bak, M., Halabowski, D., Kryk, A., Lewin, I., & Sowa, A. (2020). Mining salinisation of rivers: Its impact on diatom (bacillariophyta) assemblages, *Fottea, Olomouc*, 20(1), 1–16. 10.5507/fot.2019.010

Beltrami, M. E., Ciutti, F., Cappelletti, C., Lösch, B., Alber, R., & Ector, L. (2012). Diatoms from Alto Adige/Südtirol (Northern Italy): Characterization of assemblages and their application for biological quality assessment in the context of the water framework directive. *Hydrobiologia*, 695(1), 153–170. 10.1007/s10750-012-1194-x

Berthon, V., Alric, B., Rimet, F. & Perga, M. E. (2014). Sensitivity and responses of diatoms to climate warming in lakes heavily influenced by humans. *Freshwater Biology*. 10.1111/fwb.12380

Besse-Lototskaya, A., Verdonschot, P. F. M., Coste, M., & Van de Vijver, B. (2011). Evaluation of European diatom trophic indices. *Ecological Indicators*, 11(2), 456–467. 10.1016/j.ecolind.2010.06.017

Bonte, M. & Zwolsman, J. J. G. (2010). Climate change induced salinisation of artificial lakes in the Netherlands and consequences for drinking water production. *Water Research*, 44(15), 4411–4424. 10.1016/j.watres.2010.06.004

Busseni, G., Caputi, L., Piredda, R., Fremont, P., Hay Mele, B., Campese, L., Scalco, E., de Vargas, C., Bowler, C., d'Ovidio, F., Zingone, A., Ribera d'Alcalà, M., & Iudicone, D. (2020). Large scale patterns of marine diatom richness: Drivers and trends in a changing ocean. *Global Ecology and Biogeography*, 29(11), 1915–1928. 10.1111/geb.13161

Cantonati, M., Zorza, R., Bertoli, M., Pastorino, P., Salvi, g., Platania, g., Prearo, M., & Pizzul, E. (2021). Recent and subfossil diatom assemblages as indicators of environmental change (including fish introduction) in a high-mountain lake. *Ecological Indicators*, 125, 107603. 10.1016/j.ecolind.2021.107603

Carayon, D., Eulin-Garrigue, A., Vigouroux, R., & Delmas, F. (2020). A new multimetric index for the evaluation of water ecological quality of French Guiana streams based on benthic diatoms. *Ecological Indicators*, 113, 106248. 10.1016/j.ecolind.2020.106248

Coste, M., Boutry, S., Tison-Rosebery, J., & Delmas, F. (2009). Improvements of the biological diatom index (BDI): Description and efficiency of the new version (BDI-2006). *Ecological Indicators*, 9(4), 621–650. 10.1016/j.ecolind.2008.06.003

Cunillera-Montcusi, D., Beklioglu, M., Canedo-Arguelles, M., Jeppesen, E., Ptacnik, R., Amorim, C. A., Arnott, S. E., Berger, S. A., Brucet, S., Dugan, H. A., Gerhard, M., Horvath, Z., Langenheder, S., Nejstgaard, J. C., Reinikainen, M., Striebel, M., Urrutia-Cordero, P., Vad, C. F., Zadereev, E. & Matias, M. (2022). Freshwater salinisation: A research agenda for a saltier world. *Trends in Ecology & Evolution*, 37(5), 440–453. 10.1016/j.tree.2021.12.005

Dalu, T., & Froneman, P. W. (2016). Diatom-based water quality monitoring in Southern Africa: Challenges and future prospects. *Water SA*, 42(4), 551–559.

D'Archino, R., & Piazzi, L. (2021). Macroalgal assemblages as indicators of the ecological status of marine coastal systems: A review. *Ecological Indicators*, 129, 107835. 10.1016/j.ecolind.2021.107835

Datta, A., Marella, T. K., Tiwari, A., & Wani, S. P. (2019). The diatoms: From eutrophic indicators to mitigators. In S. K. Gupta & F. Bux (eds), *Application of Microalgae in Wastewater Treatment*, Springer Nature Switzerland, pp. 19–40. 10.1007/978-3-030-13913-1_2

Delgado, C., Pardo, I., & García, L. (2012). Diatom communities as indicators of ecological status in Mediterranean temporary streams (Balearic Islands, Spain). *Ecological Indicators*, 15(1), 131–139. 10.1016/j.ecolind.2011.09.037

Dell'uomo, A., & Torrisi, M. (2011). The eutrophication/pollution index-diatom based (EPI-D) and three new related indices for monitoring rivers: The case study of the river Potenza (the Marches, Italy). *Plant Biosystems - An International Journal Dealing with all Aspects of Plant Biology*, 145(2), 331–341. 10.1080/11263504.2011.569347

Falkowski, P. G., Barber, R. T., & Smetacek, V. (1998). Biogeochemical controls and feedbacks on ocean primary production. *Science*, 281(5374), 200–206. 10.1126/science.281.5374.200

Feio, M. J., Aguiar, F. C., Almeida, S. F. P., & Ferreira, M. T. (2012). AQUAFLORA: A predictive model based on diatoms and macrophytes for streams water quality assessment. *Ecological Indicators*, 18, 586–598. 10.1016/j.ecolind.2012.01.016

Feio, M. J., Serra, S. R. Q., Mortágua, A., Bouchez, A., Rimet, F., Vasselon, V., & Almeida, S. F. P. (2020). A taxonomy-free approach based on machine learning to assess the quality of rivers with diatoms. *Science of The Total Environment*, 722, 137900. 10.1016/j.scitotenv.2020.137900

Fernández, M. R., Martín, G., Corzo, J., de la Linde, A., García, E., López, M., & Sousa, M. (2018). Design and testing of a new diatom-based index for heavy metal pollution. *Archives of Environmental Contamination and Toxicology*, 74(1), 170–192. 10.1007/s00244-017-0409-6

Fragoso, G. M., Poulton, A. J., Yashayaev, I. M., Head, E. J. H., Johnsen, G., & Purdie, D. A. (2018). Diatom biogeography from the labrador sea revealed through a trait-based approach [Original Research]. *Frontiers in Marine Science*, 5. 10.3389/fmars.2018.00297

Fu, W., Shu, Y., Yi, Z., Pan, Y., Zhang, F., & Brynjolfsson, S. (2022). Diatom morphology and adaptation: Current progress and potentials for sustainable development. *Sustainable Horizons*, 2, 100015. 10.1016/j.horiz.2022.100015

Gómez, N., & Licursi, M. (2001). The pampean diatom index (IDP) for assessment of rivers and streams in Argentina. *Aquatic Ecology*, 35(2), 173–181. 10.1023/A:1011415209445

González-Paz, L., Delgado, C., & Pardo, I. (2020). Understanding divergences between ecological status classification systems based on diatoms. *Science of The Total Environment*, 734, 139418. 10.1016/j.scitotenv.2020.139418

Gottschalk, F., Ort, C., Scholz, R., & Nowack, B. (2011). Engineered nanomaterials in rivers–exposure scenarios for Switzerland at high spatial and temporal resolution. *Environment Pollution*, 159(12), 3439–3445. 10.1016/j.envpol.2011.08.023

Hall, R., & Smol, J. (2010). Diatoms as indicators of lake eutrophication. In J. Smol & E. Stoermer (Eds.), *The Diatoms: Applications for the Environmental and Earth Sciences*. Cambridge: Cambridge University Press, pp. 122–151. 10.1017/CBO9780511763175.008

Harding, W. R., & Taylor, J. (2011). *The South African Diatom Index (SADI): A Preliminary Index for Indicating Water Quality in Rivers and Streams in Southern Africa: Report to the Water Research Commission*. Water Research Commission.

Harper, D. (1992). The nutrients causing eutrophication, and their sources. In D. Harper (ed), *Eutrophication of Freshwaters*. Dordrecht: Springer, 1, pp. 29–59. 10.1007/978-94-011-3082-0_2

Hoek, C. van den, Mann, D. G., & Jahns, H. M. (1995). *Algae: An Introduction to Phycology*. Cambridge University Press. ISBN 0521316871.

Hong, I., Horton, B. P., Hawkes, A. D., O'Donnell, R. J., Padgett, J. S., Dura, T., & Engelhart, S. E. (2021). Diatoms of the intertidal environments of Willapa Bay, Washington, USA as a sea-level indicator. *Marine Micropaleontology*, 167, 102033. 10.1016/j.marmicro.2021.102033

Juggins, S., Kelly, M., Allot, T., Kelly-Quinn, M., & Monteith, D. (2016). A water framework directive-compatible metric for assessing acidification in UK and Irish Rivers using diatoms. *Science of Total Environment*, 568, 671–678. 10.1016/j.scitotenv.2016.02.163

Juggins, S., Kelly, M., Allott, T., Kelly-Quinn, M., & Monteith, D. (2016). A water framework directive-compatible metric for assessing acidification in UK and Irish rivers using diatoms. *Science of The Total Environment*, 568, 671–678. 10.1016/j.scitotenv.2016.02.163

Kaleli, M. A., Kociolek, J. P., & Solak, C. N. (2020). Taxonomy and distribution of diatoms on the Turkish Mediterranean Coast, Dalyan (Muğla). *Mediterranean Marine Science*, 21(1), 201–215. 10.12681/mms.17293

Kalyoncu, H., & Akkoz, C. (2009). Comparative performance of diatom indices in aquatic pollution assessment. *African Journal of Agricultural Research*, 4(10), 1032–1040.

Kang, W., Anslan, S., Börner, N., Schwarz, A., Schmidt, R., Künzel, S., Rioual, P., Echeverría-Galindo, P., Vences, M., Wang, J., & Schwalb, A. (2021). Diatom metabarcoding and microscopic analyses from sediment samples at Lake Nam Co, Tibet: The effect of sample-size and bioinformatics on the identified communities. *Ecological Indicators*, 121, 107070. 10.1016/j.ecolind.2020.107070

Kelly, M. G. (1998). Use of the trophic diatom index to monitor eutrophication in rivers. *Water Research*, 32(1), 236–242. 10.1016/S0043-1354(97)00157-7

Kelly, M. G., & Whitton, B. A. (1995). The trophic diatom index: A new index for monitoring eutrophication in rivers. *Journal of Applied Phycology*, 7(4), 433–444. 10.1007/BF00003802

Kennedy, B., & Buckley, Y. M. (2021). Use of seasonal epilithic diatom assemblages to evaluate ecological status in Irish Lakes. *Ecological Indicators*, 129, 107853. 10.1016/j.ecolind.2021.107853

Khan, M. N., & Mohammad, F. (2014). Eutrophication: Challenges and solutions. In A. A. Ansari, & S. S. Gill (eds.), *Eutrophication: Causes, Consequences and Control*. Springer Science+Business Media Dordrecht, pp. 1–15. 10.13140/2.1.3673.8884

Lavoie, I., Campeau, S., Zugic-Drakulic, N., Winter, J. G., & Fortin, C. (2014). Using diatoms to monitor stream biological integrity in Eastern Canada: An overview of 10 years of index development and ongoing challenges. *Science of The Total Environment*, 475, 187–200. 10.1016/j.scitotenv.2013.04.092

Lee, K. L., Choi, J. S., Lee, J. H., Jung, J. H., Jung, K. Y., & Kim, H. S. (2021). Response of epilithic diatom assemblages to weir construction on the Nakdong River, Republic of Korea. *Ecological Indicators*, 126, 107711. 10.1016/j.ecolind.2021.107711

Li, C. L., Ashworth, M. P., Witkowski, A., Lobban, C. S., Zglobicka, I., Kurzydlowski, K. J., & Qin, S. (2016). Ultrastructural and molecular characterization of diversity among small araphid diatoms all lacking rimoportulae. I. Five new genera, eight new species. *Journal of Phycologym*, 52(6), 1018–1036. 10.1111/jpy.12460

Li, F., Jiang, L., Zhang, T., Qiu, J., Lv, D., Su, T., Li, W., Xu, J., & Wang, H. (2021). Combined effects of seawater acidification and benzo(a)pyrene on the physiological performance of the marine bloom-forming diatom *Skeletonema costatum*. *Marine Environmental Research*, 169, 105396. 10.1016/j.marenvres.2021.105396

Li, W., Gao, K., & Beardall, J. (2012). Interactive effects of ocean acidification and nitrogen limitation on the diatom Phaeodactylum *tricornutum*. *PLOS ONE*, 8, 8. 10.1371/journal.pone.0051590

Lopez, C. L. M., & Rondon, C. A. R. (2022). Diatom response to environmental gradients in the high mountain lakes of the Colombia's Eastern range. *Aquatic Sciences*, 84, 15. 10.1007/s00027-021-00838-z

Lougheed, V., Parker, C. A., & Stevenson, R. J. (2007). Using non-linear responses of multiple taxonomic groups to establish criteria indicative of wetland biological condition. *Wetlands*, 27(1), 96–109. 10.1672/0277-5212(2007)27[96:UNROMT]2.0.CO;2

Masouras, A., Karaouzas, I., Dimitriou, E., Tsirtsis, G., & Smeti, E. (2021). Benthic diatoms in river biomonitoring—present and future perspectives within the water framework directive. *Water*, 13(4), 478.

Meyer, A., Alric, B., Dézerald, O., Billoir, E., Coulaud, R., Larras, F., Mondy, C. P., & Usseglio-Polatera, P. (2022). Linking micropollutants to trait syndromes across freshwater diatom, macroinvertebrate, and fish assemblages. *Water*, 14(8). 10.3390/w14081184

Mulholland, P. J., Webster, J. R. (2010). Nutrient dynamics in streams and the role of J-NABS. *Journal of the North American Benthological Society*, 29(1), pp. 100–117. 10.1899/08-035.1

Nixon, S. W. (2009). Eutrophication and the macroscope. *Hydrobiologia*, 629(1), 5–19. 10.1007/s10750-009-9759-z

Pandey, L. K., Sharma, Y. C., Park, J., Choi, S., Lee, H., Lyu, J., & Han, T. (2018). Evaluating features of periphytic diatom communities as biomonitoring tools in fresh, brackish and marine waters. *Aquatic Toxicology*, 194, 67–77. 10.1016/j.aquatox.2017.11.003

Pound, K. L., Larson, C. A., & Passy, S. I. (2020). Current distributions and future climate-driven changes in diatoms, insects and fish in U.S. streams. *Global Ecology and Biogeography*, 30(1), 63–78. 10.1111/geb.13193

Reid, M. A., Tibby, J. C., Penny, D., & Gell, P. A. (1995). The use of diatoms to assess past and present water quality. *Australian Journal of Ecology*, 20(1), 57–64. 10.1111/j.1442-9993.1995.tb00522.x

Rivera-Rondon, C., & Catalan, J. (2020). Diatoms as indicators of the multivariate environment of mountain lakes, *Science of The Total Environment*, 703, 135517. 10.1016/j.scitotenv.2019.135517

Round, F. E., Crawford, R. M., & Mann, D. G. (1990). *The Diatoms. Biology and Morphology of the Genera*. Cambridge University Press. ISBN 0521363187

Salmaso, F., Quadroni, S., Compare, S., Gentili, G., & Crosa, G. (2019). Benthic diatoms as bioindicators of environmental alterations in different watercourses of northern Italy. *Environmental Monitoring and Assessment*, 191(3), 158. 10.1007/s10661-019-7290-x

Simonsen, R. (1979). The diatom system: Ideas on phylogeny. *Bacillaria*, 2, 9–71.

Solak, C. N. (2011a). The application of diatom indices in the upper Porsuk Creek Kütahya-Turkey. *Turkish Journal of Fisheries and Aquatic Sciences*, 11(1). https://dergipark.org.tr/en/pub/trjfas-ayrildi/issue/13274/160358

Solak, C. N. (2011b). Water quality monitoring in European an Turkish Rivers using diatoms. *Turkish Journal of Fisheries and Aquatic Sciences*, 11(2), 329-337,10.4194/trjfas.2011.0218

Stenger-Kovács, C., Körmendi, K., Lengyel, E., Abonyi, A., Hajnal, É., Szabó, B., Buczkó, K., & Padisák, J. (2018). Expanding the trait-based concept of benthic diatoms: Development of trait- and species-based indices for conductivity as the master variable of ecological status in continental saline lakes. *Ecological Indicators*, 95, 63–74. 10.1016/j.ecolind.2018.07.026

Stevenson, J. (2014). Ecological assessments with algae: A review and synthesis. *Journal of Phycology*, 50(3), 437–461. 10.1111/jpy.12189

Szczepocka, E., & Żelazna-Wieczorek, J. (2018). Diatom biomonitoring–scientific foundations, commonly discussed issues and frequently made errors. *Oceanological and Hydrobiological Studies*, 47(3), 313–325.

Szczepocka, E., Nowicka-Krawczyk, P., & Kruk, A. (2018). Deceptive ecological status of urban streams and rivers—evidence from diatom indices. *Ecosphere*, 9(7), e02310. 10.1002/ecs2.2310

Tapolczai, K., Bouchez, A., Stenger-Kovács, C., Padisák, J., & Rimet, F. (2017). Taxonomy- or trait-based ecological assessment for tropical rivers? Case study on benthic diatoms in Mayotte island (France, Indian Ocean). *Science of The Total Environment*, 607–608, 1293–1303. 10.1016/j.scitotenv.2017.07.093

Tapolczai, K., Keck, F., Bouchez, A., Rimet, F., Kahlert, M., & Vasselon, V. (2019). Diatom DNA metabarcoding for biomonitoring: Strategies to avoid major taxonomical and bioinformatical biases limiting molecular indices capacities [Original Research]. *Frontiers in Ecology and Evolution*, 7. 10.3389/fevo.2019.00409

Tapolczai, K., Vasselon, V., Bouchez, A., Stenger-Kovács, C., Padisák, J., & Rimet, F. (2019). The impact of OTU sequence similarity threshold on diatom-based bioassessment: A case study of the rivers of Mayotte (France, Indian Ocean). *Ecology and Evolution*, 9(1), 166–179. 10.1002/ece3.4701

Teittinen, A., Virta, L. (2021). Exploring multiple aspects of taxonomic and functional diversity in microphytobenthic communities: Effects of environmental gradients and temporal changes. *Frontiers in Microbiology*, 12, 668993. 10.3389/fmicb.2021.668993

Tong, S., Xu, D., Wang, Y., Zhang, X., Li, Y., Wu, H., & Ye, N. (2020). Influence of ocean acidification on thermal reaction norms of carbon metabolism in the marine diatom *Phaeodactylum tricornutum*. *Marine Environmental Research*, 164, 105233.

Torres-Franco, A. F., Alatrista-Gongora, G. R., Guzman-Rodriguez, N. P., Calizaya-Anco, J. A., Mota, C. R., & Figueredo, C. C. (2019). Physicochemical and diatom trophic state indexes: A complementary approach for improving water sustainability in a high Andean urban stream, *Ecohydrology and Hydrobiology*, 19(4), 577–587. 10.1016/j.ecohyd.2019.01.007

Trábert, Z., Duleba, M., Bíró, T., Dobosy, P., Földi, A., Hidas, A., Kiss, K. T., Óvári, M., Takács, A., Várbíró, G., Záray, G., & Ács, É. (2020). Effect of land use on the benthic diatom community of the Danube River in the Region of Budapest. *Water*, 12(2). 10.3390/w12020479

Vasselon, V., Domaizon, I., Rimet, F., Kahlert, M., & Bouchez, A. (2017). Application of high-throughput sequencing (HTS) metabarcoding to diatom biomonitoring: Do DNA extraction methods matter? *Freshwater Science*, 36(1), 162–177. 10.1086/690649

Vasselon, V., Rimet, F., Tapolczai, K., & Bouchez, A. (2017). Assessing ecological status with diatoms DNA metabarcoding: Scaling-up on a WFD monitoring network (Mayotte island, France). *Ecological Indicators*, 82, 1–12. 10.1016/j.ecolind.2017.06.024

Vasselon, V., Rimet, F., Domaizon, I., Monnier, O., Reyjol, Y., & Bouchez, A. (2019). Assessing pollution of aquatic environments with diatoms' DNA metabarcoding: Experience and developments from France water framework directive networks. *Metabarcoding and Metagenomics*, 3, 101–115. 10.3897/mbmg.3.39646

Virta, L., Gammal, J., Jarnstrom, M., Bernard, G., Soininen, J., & Norkko, A. (2019). The diversity of benthic diatoms affects ecosystem productivity in heterogeneous coastal environments. *Ecological Society of America*, 100(9), e02765. 10.1002/ecy.2765

Wang, Y., Wu, N., Tang, T., Wang, Y., & Cai, Q. (2021). Small run-of-river hydropower dams and associated water regulation filter benthic diatom trais and affect functional diversity. *Science of Total Environment*, 813, 152566. 10.1016/j.scitotenv.2021.152566

Williams, D. (2007). Diatom phylogeny: Fossils, molecules and the extinction of evidence. *Comptes Rendus Palevol*, 6(6–7), 505–514. 10.1016/j.crpv.2007.09.016

Wingert, C. J., & Cochlan, W. P. (2021). Effects of ocean acidification on the growth, photosynthetic performance, and domoic acid production of the diatom *pseudo-nitzschia Australis* from the California current system. *Harmful Algae*, 107, 102030. 10.1016/j.hal.2021.102030

Zaghloul, A., Saber, M., Gadow, S., & Awad, F. (2020). Biological indicators for pollution detection in terrestrial and aquatic ecosystems. *Bulletin of the National Research Centre*, 44(1), 127. 10.1186/s42269-020-00385-x

Zelinka, M. (1961). Zur Prazisierung der biologischen klassifikation der Reinheid fliessender Gewasser. *Arch. Hydrobiol.*, 57, 389–407

Zhao, Q., Zhang, Y., Guo, F., Leigh, C., & Jia, X. (2020). Increasing anthropogenic salinisation leads to declines in community diversity, functional diversity and trophic links in mountain streams. *Chemosphere*, 263, 127994. 10.1016/j.chemosphere.2020.127994

4 Estuarine Diatom Community Dynamics due to Sedimentation Changes

Jelena Z. Andrejić
Department of Biological and Environmental Sciences, Georgia College & State University, Milledgeville, USA

The University of the Bahamas, School of Chemistry, Environmental and Life Sciences, Oakes Field Campus, Nassau, The Bahamas

Kalina M. Manoylov
Department of Biological and Environmental Sciences, Georgia College & State University, Milledgeville, USA

CONTENTS

4.1 INTRODUCTION

Littoral zones can be characterized with temporal exposure of algae to diurnal desiccation at low tides and suspending of microbial communities within the water column at idle high tides. These communities are of high importance for maintaining ecosystem services of the habitat (Barbier et al., 2011). Combinations of diverse freshwater, marine, and brackish diatoms dominate exposed mud samples (Manoylov and Dominy, 2013). Mud samples found within littoral zones contain both epipelic and epipsammic diatoms (Vos and Wolf, 1988). Primary producers serve as a food source for many groups of surface sediment feeders (Morrisey, 1988) and also act as agents of biogenic stabilization through the fast synthesis of a mucilaginous matrix (bio-film) during low flow. Mud samples within estuaries reflect algal distribution patterns, related to gradients of temperature, desiccation, salinity, and organic pollution (Sullivan, 1975). Therefore, the distribution of primary producers like diatoms, within a mud sample, is a result of complex interactions varying between environmental factors and inter- and intra-specific competitive interactions (Underwood and Paterson, 1993; Underwood and Provot, 2000).

DOI: 10.1201/9781003322115-4

Epipelic diatoms are the primary inhabitants of mudflats and are found within a thin horizontal layer. At the sediment-water-air interface, some of those diatoms move in a time-dependent fashion to the surface during favorable conditions (Thomson and Manoylov, 2017). Light penetration throughout the sediment is around 2–3 mm (MacIntyre et al., 1996), and the variability of pH, O_2, and nutrients can decrease exponentially within only a few millimeters of depth (Jorgensen and Revsbech, 1983). Studies have shown that migrating cells may reach 6 to 10 mm in depth within the sediment and will surface during daytime emersion and then submerge again during the night and high tides (Du et al., 2010). Epipelic diatoms also act as a major food source for many deposit and suspension feeders (MacIntyre et al., 1996). They can also play a significant role in sediment stabilization. Production of thick mucilaginous films (formed in order to survive periods of desiccation) limit solute diffusion as well as sediment erosion (Delgado et al. 1991; Holland et al. 1974).

Allometric relationships play a huge role for animals and their environment because the body shape and size of an organism determines how that organism interacts with its environment. Diatoms of different shapes and sizes are not affected the same way by physical processes such as surface adhesion, evaporation, or heat exchange. These processes determine how that organism can adapt to the environment and maintain them. Few attempts have been made to standardize the calculation of algal biovolume (Hillebrand et al., 1999). Attempts were made by Elder (1979) and Kovala and Larrance (1966) to compile geometric models for biovolume calculation. The available dimensions, the abundance and the importance of the species, and the questions to be answered should all be taken into account. Equations for benthic algal populations have been poorly researched. Elder's (1979) proposed equations for *Licmophora* and *Cymbella* were too simple, with those representatives of the genus *Ditylum,* for example, overestimated 80% using the cylinder shape formulas. A few more recent standardized calculations of algal biovolume are more accurate (Hillebrand et al., 1999).

Savannah Harbor Entrance Channel began in late 2015 (Georgia SHEP http://www.shep.uga.edu/reports.html, Georgia Port Authority, 2017), and sampling of primary producers was timed with the dredging and deepening of the port (Army Corp of Engineers, 2016). This is the only research to our knowledge relating to microbial community analyses with before and after anthropogenic influence at such a large scale. The top 50 cm water column were sampled at idle high tide (algae in those samples were called planktonic), and on the same day and same location sediment top 2 mm surface in triplicate (algae from those locations were called benthic). Objectives for the study were: 1. calculate the ratio between live and deposited (dead) diatoms; 2. document planktonic and benthic diatom proportions; and 3 differentiate changes in salinity preference for marine, brackish, and freshwater taxa within the two sampling types. Community ecology tools were used to understand diatom abilities to survive in the soft mud of the Savannah River Estuary during altered turbidity and salinity due to anthropogenic alterations of the coastal area.

4.2 MATERIALS AND METHODS

4.2.1 Study Site

The Savanna River Basin is 27,000 km^2 in eastern Georgia and western South Carolina; it defines the boundary of the two states and some of southwestern North Carolina (Dominy, 2012). The Savannah River Estuary is characterized by its unique and dynamic hydrology and complex habitats. Organisms in the system are adapted to the continuous river flow and daily tidal events. In a 24-hour and 50-min span, two high tides and two low tides occur. It takes approximately 6 hours and 12 minutes for low tide to reach high tide and vice versa (NOAA, 2017); sampling at idle point was possible during the same day. Approximately 32 km up from where the river drains into the Atlantic, saltwater begins mixing with the river's freshwater. Benthic and phytoplankton samples were collected from the Savannah River from October to December in 2015 at the Savannah River USGS site 2198920, location Lat 32009'57" Long 81009'14".

4.2.2 Collection and Processing

All sample collection followed standard protocols for sampling methods and processing (APHA, 2005). In this paper, the term "phytoplankton" was used for algae collected as grab 1L sample from the top 50 cm water column surface at high tide idle point standing at the riverbank. Composite benthic samples were taken by scraping the top 1–5 mm surface layer of the sediment with a metal frame of 5 × 5 cm and spatula. All samples were collected in triplicate for random and comprehensive representation.

4.2.3 Community Analyses

The enumeration and taxonomic identification of the algal community was based on averages. Whole-community analyses were performed in a Palmer Maloney counting chamber (colonial forms can be observed as the depth of the chamber is 0.4 mm) or on flat slides at 400x magnification (algae flattened in a single layer, higher resolution achieved). A minimum of 300 live natural algal units were classified in three categories: 1. live diatoms (unicellular or chain forms), 2. dead diatoms, and 3. all other live algal groups combined (unit represent single cells, filaments, or colonies). They were identified and enumerated to the lowest taxonomic level when possible. Then, the remainder of the subsample in the Palmer Maloney chamber was transferred on a flat slide to flattened algae in a single layer and two to five additional subsamples were scanned to observe new algal species to determine the species richness until no new live taxa were observed on two consecutive transects. Images were taken during counting and on flat slides. For samples with a low number of live cells, at least 100 natural algal units were counted. In the total community assessment, all algae analyzed were assumed to be physiologically active due to visible, healthy-looking chloroplasts (Plates 4.1 and 4.2).

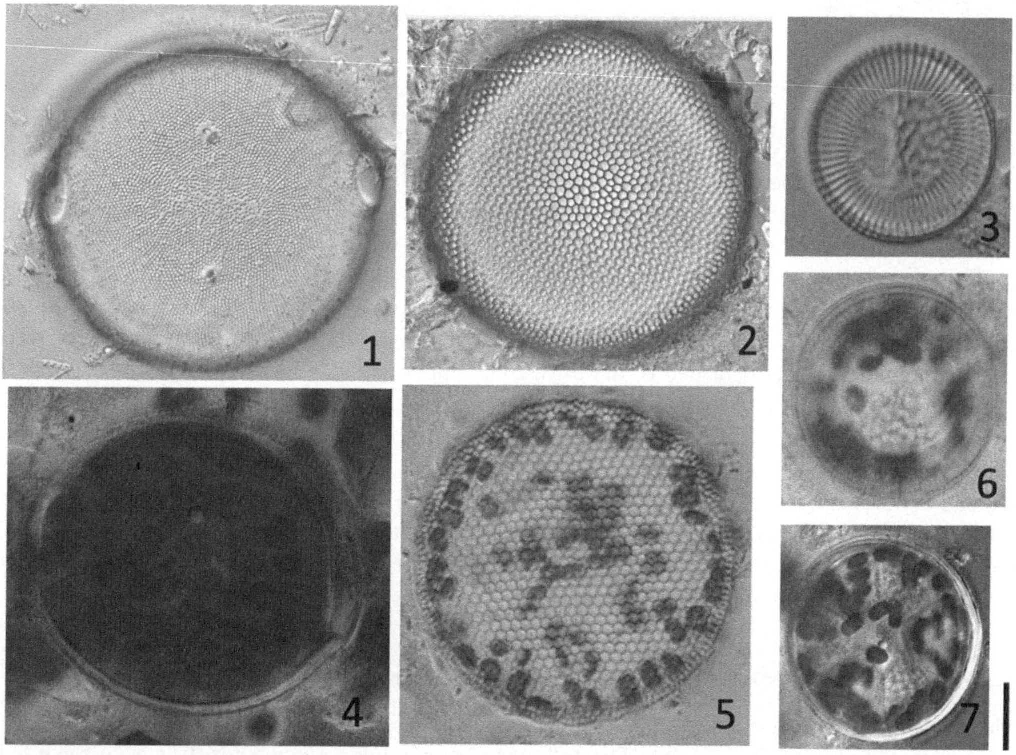

PLATE 4.1 Centric diatoms preserved live specimens from the Savannah River.

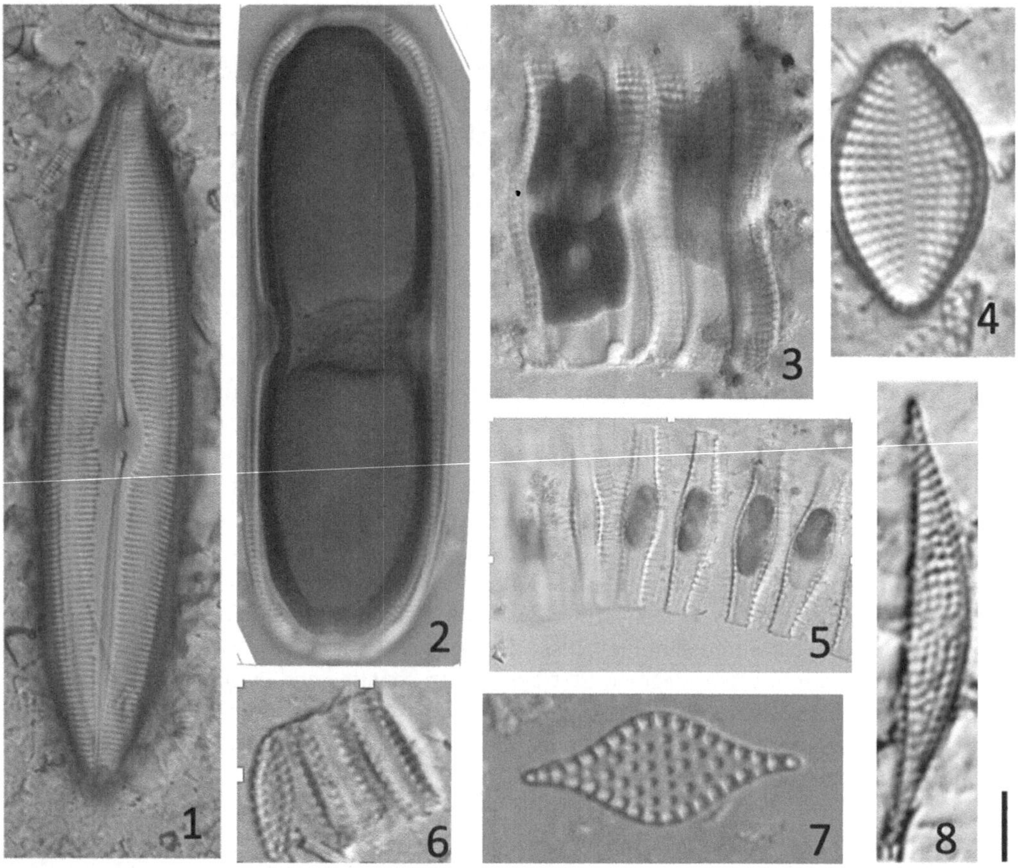

PLATE 4.2 Savannah River diatoms preserved live specimens and cleaned diatom material.

The taxonomy of the diatom community was specifically assessed on samples treated with a strong acid to increase species-level identification accuracy. 20 mL of preserved samples were digested. Permanent diatom slides were prepared with neutralized acid-cleaned material mounted in Naphrax resin for increased clarity (RI 1.74, The Biology Shop, Hazelbrook, New South. Wales, Australia). For the diatom counts, a minimum of 600 diatom valves enumerated and identified. Then, the whole surface of the cover slip was scanned, and all additional diatom species were recorded. Diatoms were identified to the lowest taxonomic level following Cremer et al. (2007), Hofmann et al., 2011, Krammer and Lange-Bertalot (1986, 1991a, 1991b), and Witkowski et al. (2000). Diatoms were classified as marine, freshwater, or brackish following published literature and algal databases (Akar and Sahin, 2017; Barnum et al., 2000; Hendey 1964; Hustedt, 1955; Lopez-Fuerte et al., 2015; Lopez-Fuerte and Siqueiros-Beltrones, 2016; Miho and Witkowski, 2005; Tsoy and Moiseenko, 2014; Vos and Wolf, 1988; Witkowski et al., 2000). Permanent slides were archived and deposited as part of the diatom slide collection of the Georgia College and State University (GCSU) Natural History Museum.

4.2.4 Biovolume and Surface-to-Volume Calculations

Measurement of biovolume: Diatoms with visible chloroplasts were considered live; species-level identification followed Cox (1996). The following morphological characteristics were recorded: length of apical axis; maximal length of transapical axis at the widest part, chloroplast shape, and number. Biovolume was calculated by approximating the shape of the diatoms to a geometric shape (Hillebrand et al., 1999). Diatoms were classified and grouped into five categories based on biovolume representing the following (Table 4.0):

TABLE 4.0

Savannah River Taxa	Ecology	Habitat	Biovolume Category
Achnanthes brevipes C. Agardh	Marine[1]	Epipelic[4]/Epiphytic[5]	3
Achnanthidium exiguum (Grunow) D.B. Czarnecki	Freshwater[1]	Epipelic[2]	1
Actinoptychus senarius (Ehrenberg) Ehrenberg	Marine[1]	Epipelic/Epiphytic[6]	4
Aulacoseira granulata (Ehrenberg) Simonsen	Freshwater[1]	Planktonic[6]	3
Bacillaria paradoxa J.F. Gmelin	Brackish[1]	Periphytic[5]	2
Caloneis cf. formosa Cleve	Freshwater	Benthic	5
Campylosira cymbelliformis (A.Schmidt) Grunow	Marine[1]	Planktonic[5]	2
Climaconeis undulata (Mei.) Lobban, Ashw. et Theriot	Marine[1]	Epipelic[4]	4
Coscinodiscus denarius A. Schmidt	Marine[1]	Planktonic	4
Cosmioneis pusilla (W.Smith) D.G. Mann & A.J. Stickle	Brackish	benthic	2
Cyclotella litoralis Lange & Syvertsen	Freshwater[1]	Planktonic[7]	4
Cylindrotheca gracilis (Brebisson ex Kutzing) Grunow	Marine[1]	Euplanktonic/Epipelic[5]	2
Cymatosira belgica Grunow	Marine[1]	Planktonic[5]	1
Cymatosira lorenziana Grunow	Marine[1]	Planktonic[5]	2
Cymbella tumida (Brébisson) van Heurck	Freshwater	Benthic	4
Decussata placenta (Ehrenberg) Lange-Bert. & Metzeltin	Freshwater	Benthic	3
Delphineis surirella (Ehrenberg) G.W. Andrews	Marine[1]	Epipsammic/Epipelic[5]	2
Diploneis aestuarii Hustedt	Marine	Epipsammic/Epipelic	1
Diploneis grundleri (A.Schmidt) Cleve	Marine	Epipsammic/Epipelic	3
Diploneis weissflogii (A.W.F. Schmidt) Cleve	Marine	Epipsammic/Epipelic	3
Encyonema turgidula Grunow	Freshwater[1]	Benthic	2
Epithemia adnata (Kützing) Brébisson	Freshwater[1]	Epipelic	2
Epithemia turgida (Ehrenberg) Kützing	Freshwater[1]	Epipelic	3
Eunotogramma marinum (W.Smith) H. Perag. & M. Peragallo	Marine[1]	Epipsammic[8]	1
Frickea lewisiana (Greville) Heiden	Brackish[1]	Epipelic[4]	5
Gomphonema parvulum (Kützing) Kützing	Freshwater[1]	Epiphytic/Epilithic[5]	1
Gomphosphenia lingulatiformis (Lan.-Bert. & Rei.) Lan.-Bert.	Freshwater[1]	Epiphytic/Epilithic[5]	1
Gyrosigma fasciola (Ehrenberg) J.W. Griffith & Henfrey	Brackish	Benthic	3
Hippodonta costulata (Grun.) L.-Bert., Met. & Witkowski	Brackish	Benthic	1
Hyalodiscus radiatus (O'Meara) Grunow	Marine	Planktonic	4
Luticola nivalis (Ehrenberg) D.G. Mann	Freshwater[1]	Benthic	2
Lyrella clavata (Gregory) D.G. Mann	Marine[1]	Benthic	3
Lyrella hennedyi (W.Smith) Stickle & D.G. Mann	Marine[1]	Benthic	4
Navicula luciae A.Witkowski & Lange-Bertalot	Marine[1]	Epipelic	2
Navicula platyventris Meister	Marine[1]	Epipelic/Planktonic[5]	1
Navicula ramosissima (Agardh) Cleve	Marine[1]	Epipelic/Planktonic[5]	2
Navicula salinarum Grunow	Brackish[1]	Epipsammic/Epipelic[5]	2
Neodelphineis silenda (Hohn & Hell.) Desianti & Potapova	Marine[1]	Planktonic[4]	1
Nitzschia brevissima Grunow	Freshwater[1]	Epipelic[5]	1
Nitzschia clausii Hantzsch	Freshwater[1]	Epipelic[5]	1

(Continued)

TABLE 4.0 (Continued)

Savannah River Taxa	Ecology	Habitat	Biovolume Category
Nitzschia prolongata var. *hoehnkii* (Hustedt) Lange-Bertalot	Brackish[1]	Epipelic[5]	2
Nitzschia scalpelliformis Grunow	Brackish[1]	Epipelic[5]	4
Nitzschia sigma (Kützing) W. Smith	Brackish[1]	Epipelic[5]	4
Odentella aurita (Lyngbye) C. Agardh	Marine[1]	Planktonic/Epiphytic[5]	4
Paralia sulcata (Ehrenberg) Cleve	Marine[1]	Planktonic/Epiphytic[5]	3
Pinnularia gibba (Ehrenberg) Ehr.	Freshwater[1]	Epipelic[4]	3
Plagiogramma pulchellum Greville	Marine[1]	Planktonic	2
Plagiogramma tessellatum Greville	Marine[1]	Planktonic	3
Planothidium apiculatum (R.M. Patrick) Lange-Bert.	Freshwater[1]	Benthic	2
Pleurosigma diverse-striatum Meister	Marine[1]	Benthic	3
Pleurosira laevis (Ehrenberg) Compère	Brackish	Planktonic/Epiphytic[5]	5
Psammodictyon panduriforme (W. Gregory) D.G. Mann	Marine[1]	Epiphytic/Epilithic[5]	2
Pseudosolenia calcar-avis (Schultze) Sundström	Marine[1]	Planktonic/Epiphytic[5]	4
Rhaphoneis amphiceros (Ehrenberg) Ehr.	Marine[1]	Planktonic[5]	3
Rhopalodia acuminata Krammer	Brackish	Epipelic[3]	3
Rhopalodia gibba (Ehrenberg) Otto Müller	Freshwater[1]	Epipelic[3]	5
Skeletonema costatum (Greville) Cleve	Marine[1]	Planktonic[5]	1
Stauroneis smithii Grunow	Freshwater[1]	Epipelic[3]	2
Surirella febigerii Lewis	Brackish[1]	Epipelic[3]	5
Tabularia fasciculata (C.Agardh) D.M. Williams & Round	Brackish[1]	Epipelic/Epilithic	1
Tabularia tabulata (C.Agardh) Snoeijs	Brackish[1]	Epipelic/Epilithic	3
Thalassiosira oestrupii var. *venrickiae* G.Fryxell & Hasle	Marine[1]	Planktonic	1
Triceratium reticulum Ehrenberg	Marine[1]	Planktonic	3
Tropidoneis seriata Cleve	Marine[1]	Epipsammic/Epipelic[5]	5
Tryblionella granulata (Grunow) D.G. Mann	Marine[1]	Epipsammic/Epipelic[5]	2

Diatom species composition from the Savannah River Estuary 2015. Genus-based biovolume category as described in Materials and Methods section. Ecology and habitat information sources: [1]AlgaeBase 2022; [2]Akar and Sahin 2017; [3]Barnum et al. 2000; [4]Lopez-Fuerte et al. 2015; [5]Lopez-Fuerte and Siqueiros-Beltrones 2016; [6]Miho and Witkowski 2005; [7]Tsoy and Moiseenko 2014; [8]Amspoker 2016

1. chain-forming araphid, some centric, monoraphid diatoms: 10–500 μm^3;
2. small Navicula, Delphineis, tube-forming Nitzschia, Psammodictyon: 501–2,500 μm^3;
3. Gyrosigma, Diploneis, Aulacoseira: 2,501–5,000 μm^3;
4. motile Nitzschia, Pseudosolenia, large Navicula, large Centric diatoms: 5,001–10,000 μm^3;
5. Pleurosira, Frickea, Tropidoneis, large Surirella: >10,001 μm^3.

Diatom abundance is expressed as relative abundance percent of the total valves counted. ANOVA was used to test the significance of patterns observed for classification of habitat preference between species (i.e., benthic or planktonic), as well as salinity preferences based on type (i.e., brackish, marine, or freshwater). All graphs and statistical analyses were performed with Systat 13, 2017 Systat Software, Inc.

4.3 RESULTS AND DISCUSSION

Physical and chemical field data for each month were analyzed based on USGC long-term site 2198920 data (Table 4.1) Air and water temperatures followed seasonal variations, pH, DO-dissolved oxygen, and pH did not vary significantly, whereas conductivity declined in November

TABLE 4.1
Physical and Chemical Field Data for 2015 Samples from the Savannah River USGC Long-term Monitoring Site 2198920; DO-dissolved Oxygen, (Note: Conductivity Represents Specific Conductance, Water, Unfiltered, (µS/cm at 25°C)

Month	Temp. (°C) (air)	Temp. (°C)	DO mg/L	pH	Conductivity (µS/cm)
Sept	31.2	23	6.2	6.8	14500
Oct	22	21	7.2	7.3	10300
Nov	17.9	15	7.8	6.9	6500
Dec	16.2	17	7.7	6.8	4000
Average ± SE	22 ± 3.3	19 ± 1.8	7.2± 0.3	6.95± 0.1	8825± 2292

and December. (Note: Conductivity represents specific conductance, water, unfiltered, µS/cm at 25°C, Table 4.1.) Turbidity increased above 350 FNU in October (immediately after dredging began) and then lowered to an average below 45 FNU. Beginning of the construction corresponded with very high salinity that stayed through October (Figure 4.1), but no data were provided for November and December.

4.3.1 Algal Community Structure

The live algal community was dominated by diatoms (Bacillariophyceae). In the phytoplanton community, live diatom percentages declined in October and November, but recovered to almost 100% in December. With the exception of November, the percentage of dead frustules in benthic samples was more than 50%. This may be due to sedimentation of the cells, and the fact that the frustules in the intertidal estuarine zone are brought up from the Savannah River and the ocean. The proportion of dead diatoms differed through time but stabilized, and almost all planktonic diatoms identified in December were alive. Proportions of live diatoms increased from November to Devember in benthic samples (Table 4.2).

During this study, water temperature, DO, and pH did not vary compared to the significant decline in conductivity and turbidity. Measured turbidity at the site increased in October and stabilized for

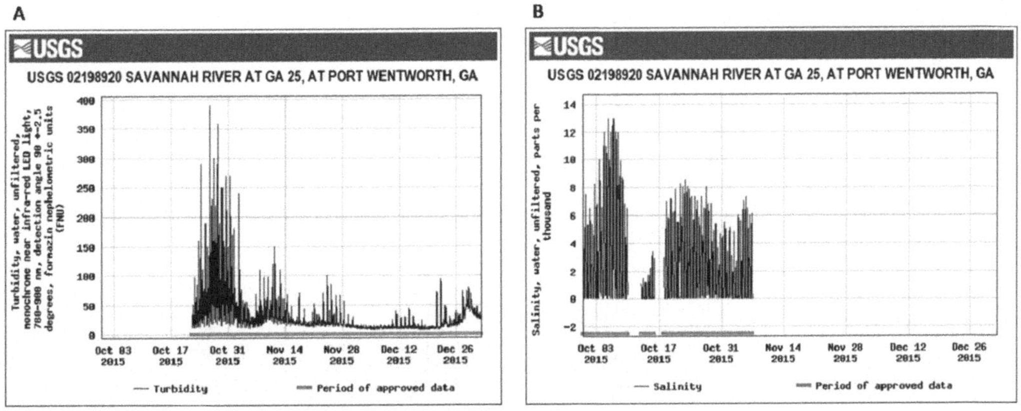

FIGURE 4.1 Physicochemical data at the Savannah River.

TABLE 4.2
Algal Community Analyses Through Time During Construction October to December 2015

Analyses	Percent Live Diatoms		Algal Species Richness		Diatom Species Richness	
Month	**Planktonic**	**Benthic**	**Planktonic**	**Benthic**	**Planktonic**	**Benthic**
Sept	80	43	33	45	72	45
Oct	72	49	40	36	50	49
Nov	42	65	47	29	42	44
Dec	98	86	66	34	54	67

November and December. Salinity was high in September and October and played a significant role in total species composition observed. In whole community analyses, phytoplankton had the highest species richness in December, but benthic species richness gradually decreased from September to December. In clean diatom analyses, phytoplankton species richness was highest in September and stayed around 50 diatom for the rest of the sampling period. Benthic diatom community richness was highest in December.

Compared with research from the same location, brackish benthic diatoms were physiologically active in all habitats and through time (Figure 4.2). Biomass was dominated by brackish diatoms. Small freshwater diatoms were replaced by very small marine diatoms, whereas freshwater medium-size diatoms remained in the communities when compared between 2011 and 2015 (Figure 4.2). Marine planktonic dominated the community, potentially due to sedimentation. Physiologically active diatom biomass decreased for all size categories and all habitats. Diatoms dominated the size distribution of those diatoms containing lipid droplets and the frequency numbers above the highest potential average of oil content within that category (Figure 4.3). Benthic diatoms contribution to the primary producer community of the smallest and largest diatoms from 2011 were lost. Small and large benthic brackish diatoms were replaced in 2015 with small to medium marine taxa.

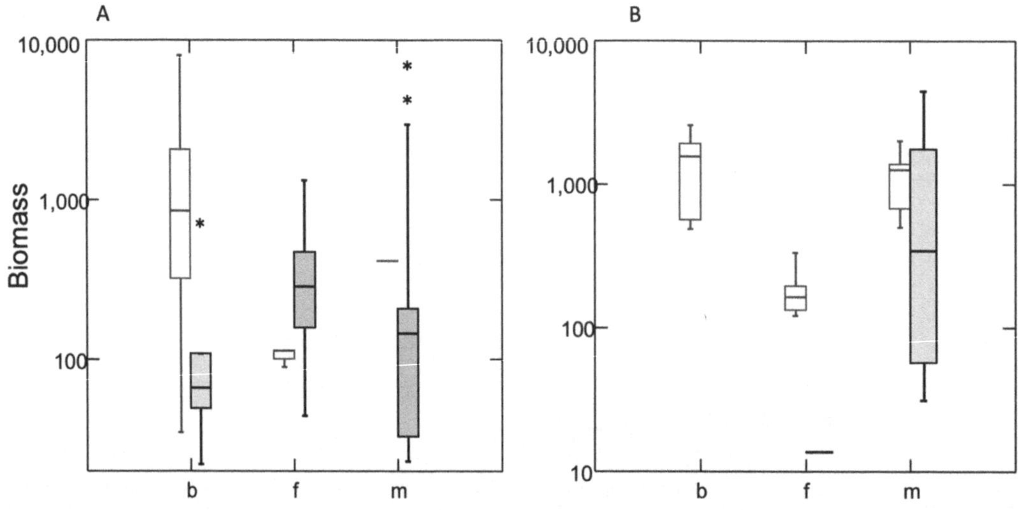

FIGURE 4.2 Diatom biomass estimations per habitat.

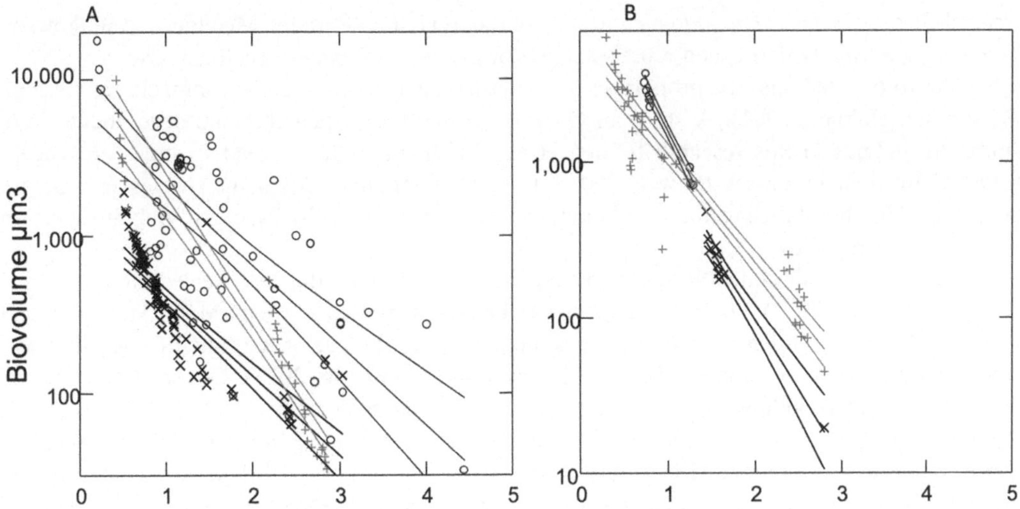

FIGURE 4.3 Diatom biovolume per habitat in five biovolume categories.

4.3.2 Community Dynamics Studies

A total of 90 diatoms were documented during this study; 65 of those taxa were identified to the species level (Supplemental Table 4S). Sampling during port construction in the Savannah River and species richness in live community analyses varied from 33 to 66 (average 47 ±7 SE) for phytoplankton and was significantly different from the bentic primary producer community 29 to 45 (average 36 ±3 SE)

For the clean diatom analyses, similar to findings in the 2011 collection from the location (Dominy, 2012), phytoplankton were 42 to 72 (average 55 ±6 SE) and benthic siatom community were 44 to 67 (average 51 ±5 SE). Species richness numbers are lower than others reported from the area (Cremer et al., 2007; Hustedt, 1955; Manoylov and Dominy, 2013). *Minidiscus sp.*, a common planctonic diatom in the 2011, collection was not documented in 2015, potentially being very sensitive to the altered water clarity and change in turbidity with anthropogenic alterations. In estuarine communities, taxonomic variation in diatoms can be explained by sediment composition and salinity (Hassan, 2008). Species-level identification remains of great importance (Manoylov, 2014), especially in centric diatoms (Plate 4.1). Salinity plays a major role in plankton biodiversity (Farhadian et al., 2015) and has been addressed for a community in the southeastern United States (Vernberg et al., 1992). Partial data on salinity were available from the USGS, but diatom brackish taxa decreased, freshwater stayed in low abundance, and marine diatoms proliferated by the end of 2015. *Cymatosira belgica* (Plate 4.2, fig. 6), *Cymatosira lorenziana* Grunow (Plate 4.2, figs 5, 7), *Delphineis surirella* (Ehrenberg) G.W. Andrews (Plate 4.2, fig. 4), and *Skeletonema costatum* (Greville) Cleve were dominant species in phytoplankton samples. Species that form chains were also present: *Chaetoceros sp.* and *Aulacoseira granulata* (Ehrenberg) Simonsen. In December, abundant marine planktonic diatoms, including *Campylosira cymbelliformis* (Schmidt) Grunow (Plate 4.1, fig. 8), *Cylindrotheca gracilis* (Brébisson) Grunow, and *Thalassionema nitzschioides* (Grunow) Mereschkowsky, increased in abundance. Pleurosira laevis (Ehrenberg) Compère (Plate 4.1, figs. 1, 4) decreased in this month.

Marine benthic diatoms dominated the community and were represented by *Achnanthes brevipes* (Plate 4.2, fig. 3), *Tryblionella littoralis* (Grunow) Mann, and *Gyrosigma litorale* (W. Smith) Cleve. Riverine freshwater diatoms, such as *Caloneis cf. formosa* (Plate 4.2, fig. 1), *Bacillaria paradoxa* Gmelin, and many representatives of the genus Navicula were potentially deposited in the mud too deep and did not move upward to the surface, where we sampled (Thomson and Manoylov, 2017).

Brackish diatoms, such as *Hippodonta costulata* (Grunow) Lange-Bertalot, Meteltin, and Witkowski, *Navicula salinarum* Grunow, and representatives of the genus Tabularia declined, also.

Similar to our findings, the proportion of dead diatoms in estuarine environments can be high (Wilson and Holmes, 1981). A study in Oregon streams shows that live and dead counts were similar to findings in this research (Gillett et al., 2009). In addition, dead frustules are widely disturbed by tidal processes (Sawai, 2001). Loss of community-dominant species is a metric incorporated in littoral management and monitoring (Taljaard et al., 2013), classifying this location of the river as impacted.

At this location, subsequent year-round sampling of primary producers revealed high microbial diversity (Dominy, 2012), and diatom species richness was previously reported in Manoylov and Dominy (2013), but information on the physiological state of algae at the time of collection was studied only in terms of motile diatoms (Thomson and Manoylov, 2017). Our methods allowed obtaining an understanding of the ecology of algae in the sediments (Manoylov et al., 2016). Population changes in primary producers will remain important approaches in understanding complex coastal communities (Desrosiers et al., 2013). The majority of diatoms in the sediment were not physiologically active (i.e., chloroplasts not present), but they were potentially deposited as silica cell wall particles. Tidal resuspension is a significant factor in determining the spatial distribution of algae (Johnson et al., 1974; Kromkamp et al., 2006). The process affects the horizontal distribution of both benthic and planktonic algae by causing shoreward transport of the cells to the mudflats, where an extensive number of empty frustules are deposited (Baillie and Welsh, 1980; Facca and Sfriso, 2007).

4.3.3 Conclusions and Perspectives

Algal identification and community-based bioassessment remain important research. Marine biofilms are diverse and complex and can contain a high number of empty valves belonging to both phytoplankton and benthos. Because traditional methods used to process diatom samples (removing the organic matter with acids) for diatom analyses do not allow differentiation of live (physiologically active) vs. already empty (dead) diatoms, there is the potential of an overestimation of species richness and diversity. In this research, studying widening and deepening of the estuarine system were documented upstream in the Savannah River that at a tidal location, diatoms responded to changes in turbidity, water clarity, and sedimentation.

This research documented phytoplankton survivorship was higher when turbidity stabilized and potentially was driven by a larger proportion of diatoms surviving in colder temperatures. The live phytoplankton community in October was dominated by *Skeletonema costatum* in October. With the start of deepening and dredging of the Savannah River Port, phytoplankton diatoms changed proportions, as the uplifting of *Pleurosira laevis* and representatives of the genus *Nitzschia* were detected. *Cymatosira belgica* dominated the benthic community in this site in 2011 and was replaced by representatives of the genus *Navicula*. Microscopic primary producer communities dominate estuarine diatom communities and are a complex mix of freshwater, marine, and brackish representatives. This research also shows the value of population dynamics based on biovolume responses in different habitats, in need of both fine-level taxonomic and community research. Deposition of marine planktonic species, loss of larger centric and motile diatoms, and a high amount of dead diatom frustules were documented within the benthic soft mudflats.

4.4 ACKNOWLEDGEMENT

The authors would like to thank two anonymous reviewers and the editor Dr Archana for the helpful suggestions. Dr. Andrejic's work was supported by Fulbright award postdoc funding to Dr. Manoylov. We are grateful to the many phycology lab student members for help with field collection and processing.

REFERENCES

APHA 2005. *Standard Methods for Examination of Water and Wastewater, American Public Health Association*. Washington DC.

Admiraal, W., and H. Peletier. 1979 Distribution of diatom species on an estuarine mudflat and experimental analysis of the selective effect of stress. *Journal of Experimental Marine Biology and Ecology* 46:157–175.

Admiraal, W., and Peletier, H. (1979). Influence of organic compounds and light limitation on the growth rate of estuarine benthic diatoms. *British Phycological Journal*, 14, 197–206. 10.1080/00071617900650211

Akar, B., and B. Sahin. 2017. Diversity and Ecology of Benthic Diatoms in Karagol Lake in Karagol-Sahara National Park (Savsat, Artvin, Turkey). *Turkish Journal of Fisheries and Aquatic Sciences* 17:15–24.

AlgaeBase 2022. Database of information on algae that includes terrestrial, marine, and freshwater organisms. http://www.algaebase.org (accessed May 15th 2022).

Amspoker, M. C. 2016. *Eunotogramma litorale* sp. nov., a marine epipsammic diatom from Southern California, USA. *Diatom Research* 31(4):389–395. 10.1080/0269249X.2016.1256350

Army Corp of Engineers 2016. http://www.sas.usace.army.mil/Missions/Civil-Works/Savannah-Harbor-Expansion/What-is-SHEP/

Baillie, P., and B. Welsh. 1980. The effect of tidal resuspension on the distribution of intertidal epipelic algae in an estuary. *Estuarine and Coastal Marine Science* 10:165–180.

Barbier, E., Hacker, S., Kennedy, C., Koch, E., Stier, A., and B. Silliman. 2011. The value of estuarine and coastal ecosystem services. *Ecological Monographs* 81:169–193.

Barnum, D., Elder, J., Stephens, D., and M. Friend. 2000. Developments in hydrobiology: The Salton Sea. *Hydrobiologia* 473:1–293.

Cox, E. 1996. *Identification of Freshwater Diatoms from Live Material*. Springer, 168 pp.

Cremer, H., Sangiorgi, F., Wagner-Cremer, F., McGee, V., Lotter, A., and H. Visscher. 2007. Diatoms (Bacillariophyceae) and Dinoflagellate Cysts (Dinophyceae) from Rookery Bay, Florida, U.S.A. *Caribbean Journal of Sciences* 43:23–58.

Desrosiers, C., Leflaive, J., Eulin, A., and L. Ten-Hage. 2013. Bioindicators in marine waters: Benthic diatoms as a tool to assess water quality from eutrophic to oligotrophic coastal ecosystems. *Ecological Indicators* 32:25–34.

Delgado, M., De Jonge, V.N., & Peletier, H. (1991). Effect of sand movement on the growth of benthic diatoms. Journal of Experimental Marine Biology and Ecology, 145, 221–231.

Dominy, J. 2012. *Algal population dynamics within the Savannah River Estuary*. Master's Thesis. Georgia College and State University.

Du, G. Y., Son, M., An, S., and I. K. Chung. 2010. Temporal variation in the vertical distribution of microphytobenthos in intertidal flats of the Nakdong River estuary, Korea. *Estuarine Coastal and Shelf Science* 86:62–70.

Elder, L. 1979. Recommendations for marine biological studies in the Baltic Sea: phytoplankton and chlorophyll. *The Baltic Marine Biologists Publ* #5:38.

Facca, C., and A. Sfriso. 2007. Epipelic diatom spatial and temporal distribution and relationship with the main environmental parameters in coastal waters. *Estuarine Coastal and Shelf Science* 75:35–49.

Farhadian, O., Pouladi, M., Vazirizadeh, A., and R. Sedaghat. 2015. A study of diatoms seasonal distribution and biodiversity in helleh river estuary. Persian Gulf. *Environmental Studies of Persian Gulf* 2:32–44.

Georgia Port Authority. 2017. http://www.gaports.com/PortofSavannah.aspx

Gillett, N., Pan, Y. and C. Parker. 2009. Should only live diatoms be used in the bioassessment of small mountain streams? *Hydrobiologia* 620:135–147. 10.1007/s10750-008-9624-5

G.T.B,1965. An Introductory Account of the Smaller Algae of British Coastal Waters. Part V. Bacillariophyceae (Diatoms) By N. I. Hendey Fishery Investigations, Series IV, Her Majesty's Stationery Office, London. 1964. Pp. xxii + 317, Plates I–XLV. Price £9. Journal of the Marine Biological Association of the United Kingdom, 45, 798–798.

Hassan, G., Espinosa, M., and F. Isla. 2008. Fidelity of dead diatom assemblages in estuarine sediments: how much environmental information is preserved? *Palaios* 23:112–120.

Hillebrand, H., Durselen, C., Kirschtel, D., Pollingher, U., and T. Zohary. 1999. Biovolume calculation for pelagic and benthic microalgae. *Journal of Phycology* 35:403–424.

Hofmann, G., M. Werum, and H. Lange-Bertalot. 2011. Diatomeen im SuBwasser-Benthos von Mitteleuropa. Bestimmungsflora Kieselalgen fur die okologische Praxis. Uber 700 haufigsten Srten und ihre Okologie. A.R.G. Gantner Verglag K.G.

Hustedt, F. 1955. Marine littoral diatoms of beaufort. *North Carolina. Duke University Marine Station Bulletin* 6:1–67.

Holland, A. F., Zingmark, R. G., & Dean, J. M. 1974. Quantitative evidence concerning the stabilization of sediments by marine benthic diatoms. Marine Biology, 27, 191–19610.1007/bf00391943

Johnson, A., Hillestad, H., Shanholtzer, S., and G. Shanholtzer. 1974. An Ecological Survey of the Coastal Region of Georgia. *National Park Service Scientific Monograph* 3:1–100.

Jorgensen, B. B., Revsbech, N. P., & Cohen, Y. 1983. Photosynthesis and structure of benthic microbial mats: Microelectrode and SEM studies of four cyanobacterial communities1. *Limnology and Oceanography*, 28, 1075–1093.

Kovala, P. E. and J.P. Larrance . 1966. Computation of phytoplankton cell numbers, cell volume, cell surface area and plasma volume per litre, from microscopical counts. Special report, vol. 38, University of Washington, Seattle, pp. 1–91.

Krammer, K. and H. Lange-Bertalot. 1986. Bacillariophyceae. 1. Teil: Naviculaceae. In *Susswasserflora von Mitteleuropa* eds.Ettl, H., Gerloff, J., Heynigh, H., and Mollenhaure, D., 1–876.

Krammer, K. and H. Lange-Bertalot. 1998a. Bacillariophyceae, Epithemiaceae, Surirellaceae. In *Susswasserflora von Mitteleuropa* eds.Ettl, H., Gerloff, J., Heynigh, H., and Mollenhaure, D., 1–596.

Krammer, K., and H. Lange-Bertalot. 1998b. Bacillariophyceae. 3. Teil: Centrales, Fragilariaceae, Eunotiaceae. In *Susswasserflora von Mitteleuropa* eds.Ettl, H., Gerloff, J., Heynigh, H., and Mollenhaure, D., 1–576, Gustav Fisher Verlag, Stuttgart, Germany.

Kromkamp, J., deBrouwer, J., Blanchard, G., Forster, R., and V. Creach. 2006. Functioning of microphytobenthos in estuaries. Proceedings of the Colloquium, Amsterdam.

Lopez-Fuerte, F., and D. Siqueiros-Beltrones. 2016. A checklist of marine benthic diatoms (Bacillariophyta) from Mexico. *Phytotaxa* 3:201–258.

Laws R.A., Diatoms (Bacillariophyceae) from surface sediments in the San Francisco Bay Estuary. 1988, 45., Proceedings of the California Academy of Sciences, 133–254.

Lopez-Fuerte, F., Siqueiros-Beltrones, D., and R. Yabur. 2015. First record of benthic diatoms (Bacillariophyceae and Fragilariphyceae) from Isla Guadalupe, Baja California, Mexico. *Revista Mexicana de Biodiversidad* 86:281–292.

MacIntyre, H., Geider, R., and D. C. Miller. 1996. Microphytobenthos: The ecological role of the “Secret Garden” of unvegetated, shallow-water marine habitats. I. Distribution, abundance and primary production. *Estuaries* 19:186–201.

Manoylov, K. 2014. Taxonomic identification of algae (morphological and molecular): species concepts, methodologies, and their implications for ecological bioassessment. *Journal of Phycology* 50:409–424.

Manoylov, K., and J. Dominy. 2013. Changes in Epipelic Diatom Diversity from the Savannah River Estuary. *Journal of Environmental Protection* 4:172–179.

Manoylov, K., France, E., Geletu A., and J. Dominy. 2016. Algal Community Membership of Estuarine Mudflats from the Savannah River, United States. *Journal of Marine Science and Engineering* 4(11) (February):1–15. www.mdpi.com/journal/jmse, doi:10.3390/jmse4010011

Miho, A.. andA. Witkowski. 2005. Diatom (Bacillariophyta) Flora of Albanian Coastal Wetlands Taxonomy and Ecology: A Review. *Proceedings of the California Academy of Sciences* 56:129–145.

NOAA. 2017. https://oceanservice.noaa.gov/education/kits/tides/tides05_lunarday.html

Morrisey, Donald J. 1988. Differences in effects of grazing by deposit-feeders Hydrobia ulvae (Pennant) (Gastropoda : Prosobranchia) and Corophium arenarium Crawford (Amphipoda) on sediment microalgal populations. II. Qualitative differences. *Journal of Experimental Marine Biology and Ecology*, 118(1), 43–53.

Palmer, C. M. and T. E. Maloney. 1954. *A New Counting Slide for Nanoplankton*. American Society of Limnology and Oceanography Special Publication, Waco, TX, USA.

Sawai, Y. 2001. Distribution of living and dead diatoms in tidal wetlands of northern Japan: relations to taxonomy. *Paleogeography* 173:125–141.

Sullivan, M. 1975. Diatom communities from a Delaware saltmarsh. *Journal of Phycology* 11:384–390.

Taljaard, S., Weerts, S. P., Whitfield, A. K., and T. H. Wooldridge. 2013. Country-wide assessment of estuary health: An approach for integrating pressures and ecosystem response in a data limited environment. *Estuarine, Coastal and Shelf Science* 130:239–251.

Thomson, A., and K. Manoylov. 2017. Benthic diatom motility in estuarine mudflats and coastal sands of the Atlantic Coast. *International Journal of Marine Biology* 2:1–9.

Thomson, A., and K. Manoylov. 2019. Algal Community Dynamics Within the Savannah River Estuary, Georgia Under Anthropogenic Stress. *Estuaries and Coasts* 42:1459–1474 10.1007/s12237-019-00579-1

Tsoy, I., and I. Moiseenko, 2014. Diatoms from surface sediments of Amurskiy Bay, Sea of Japan. *Russian Journal of Marine Biology* 40:10–23.

Underwood, G., and D. Paterson. 1993. Seasonal changes in diatom biomass, sediment stability and biogenic stabilization in the Severn Estuary. *Journal of Marine Biology* 73:871–887.

Underwood, G., and L. Provot. 2000. Determining the environmental preferences of four estuarine epipelic diatom taxa: growth across a range of salinity, nitrate and ammonium conditions. *European Journal of Phycology* 35:173–182.

Vernberg, F., Vernberg, W., Blood, E., Fortner, A., Fulton, M., McKellar, H., Michener, W. Scott, G., Siewicki, T., and K. Figi. 1992. Impact of Urbanization on High-Salinity Estuaries in Southeastern United States. *Netherlands Journal of Sea Research* 30:239–248.

Vos, P. C., and H. de Wolf. 1988. Methodological aspects of paleo-ecological diatom research in coastal areas of the Netherlands. *Geologic en Mijinbouw* 67:31–40.

Wilson, C., and R. Holmes. 1981. The ecological importance of distinguishing between living and dead diatoms in estuarine sediments. *Journal of Phycology* 16:345–349.

Witkowski, A., Lange-Bertalot, H., and D. Metzeltin. 2000. Diatom Flora of Marine Coasts I. In. *Iconographia DIatomologica, Annotated Diatom Micrographs*, ed.H. Lange Bertalot, 1:1–925. A.R.G. Gantner Verlag Press.

5 Carbon Dioxide Mitigation by Diatoms

Mohd Azwan Ahmad
Department of Diagnostic and Allied Health Science, Faculty of Health and Life Sciences, Management and Science University, Selangor, Malaysia

Murni Amirra Mohd Aminuddin
Centre of Environmental Health and Safety, Faculty of Health Sciences, UiTM Cawangan Selangor Kampus Puncak Alam, Puncak Alam, Selangor, Malaysia

Nur Faiza Abdul Razak
Technical Foundation Section, Institute of Medical Science Technology, Universiti Kuala Lumpur, Kajang, Selangor, Malaysia

Wai Yan Cheah
Centre of Research in Development, Social and Environment (SEEDS), Faculty of Social Sciences and Humanities, Universiti Kebangsaan Malaysia, UKM Bangi, Selangor Darul Ehsan, Malaysia

CONTENTS

5.1 INTRODUCTION – CARBON DIOXIDE GENERATION AND EMISSION

Diatoms are the major group of algae and are among the most common types of phytoplankton. Diatoms are unicellular organisms, although they can form colonies. Diatoms are unicellular microalgae possessing a silicon-based cell wall. Diatoms belong to the kingdom of protista. At least 100,000 species of diatoms have been discovered. Diatoms exist in a large number in most bodies of water throughout the world. In fact, just one liter of seawater can contain one million diatoms. They are primary producers and create a very large food source in the ocean. Diatoms are autotrophs and produce energy by photosynthesis. As a result, diatoms need sunlight,

DOI: 10.1201/9781003322115-5

carbon dioxide (CO_2), and water. They are found in the ocean, freshwater, and even moist topsoil. Many diatoms are microscopic, but some measure a millimeter across. Diatoms contribute an enormous amount of oxygen to our atmosphere. Diatoms are organized into different orders if they are centrales or pennales. Centrales have radial symmetry and are most often found in the ocean. They are wheel shaped. They can be found drifting near the surface, absorbing sunlight and other nutrients. Pennales tend to be elongated. Many pennales live in freshwater streams, in swamps, in ditches, or on the bottom of shallow regions of oceans and estuaries. They are one of the few organisms that can live in the open ocean, polar seas, tropical waters, all freshwater locations, dirt, snow, and even glacial ice. However, various diatom species must acquire distinct adaptations to survive in diverse environments (Jordan Ramsby et al., 2020).

CO_2 is the primary greenhouse gas emitted through human activities. Human activities are altering the carbon cycle – both by adding more CO_2 to the atmosphere and by influencing the ability of natural sinks, like forests and soils, to remove and store CO_2 from the atmosphere. While CO_2 emissions come from a variety of natural sources, human-related emissions are responsible for the increase that has occurred in the atmosphere since the Industrial Revolution (Parry, 2015). Reduced fossil fuel consumption is the most effective approach to minimize CO_2 emissions. Many CO_2 reduction measures are cross-cutting and can be applied to homes, businesses, industry, and transportation. Adding building insulation, driving more fuel-efficient automobiles, and using more energy-efficient electrical equipment are all strategies to cut down on energy consumption and consequently CO_2 emissions. Personal energy use could be reduced by turning off lights and electronics when they are not in use. Reduced vehicle distance travelled and minimized gasoline usage are methods of conserving energy and lowering CO_2 emissions. Carbon emissions can be reduced by producing more energy from renewable sources and using fuels with lower carbon content. CO_2 capture and sequestration are also related technologies that can potentially reduce CO_2 emissions from new and existing coal- and gas-fired power plants, industrial processes, and other stationary sources of CO_2. Measures to mitigate the emissions of CO_2 can vary substantially in terms of the energy required. Some suggested CO_2 reduction approaches require energy-intensive processes that jeopardize their viability as mitigation options, particularly if implemented globally (Babacan et al., 2020). CO_2 mitigation by photosynthetic organisms can be a sustainable alternative when coupled with bioprocessing and biomanufacturing for added value. Therefore, the potential of diatoms for CO_2 mitigation, coupled with bio-based manufacturing, highlights the challenges to be overcome for a commercially viable, sustainable manufacturing solution to minimize CO_2 emissions.

The photosynthetic production of molecular oxygen, otherwise known as oxygenic photosynthesis, was observed previously (Sethi et al., 2020). Diatoms are one of the few technologies using CO_2 generated by human activities on a global scale. Diatoms, a natural CO_2 sink, can be grown in an industrial setting in both autotrophic and mixotrophic modes for CO_2 mitigation. Diatoms are a group of globally distributed phytoplankton, with an estimated contribution of 25% to global primary production. (Bach et al., 2019). They have evolved from their ancestors, from about 250 to 190 million years ago (Medlin, 2016), have become a highly diverse and biogeochemically relevant group of phytoplankton, and contribute significantly to the natural carbon sink.

They are a reservoir of untapped potential with a wide range of applications, including CO_2 reduction, primary producers in the aquatic food web, and wastewater clean-up by quenching toxins from a variety of sources, including industries, agriculture, and human sources. The marine diatoms, as a group, have the potential for high CO_2 fixing capacity, being naturally evolved for this function. However, further research into carbon fixation is needed before industrial-scale solutions can be devised to maximize carbon intake and route fixed carbon to industrial-scale operations. Diatoms' evolutionary success story began only 200 million years ago, but they have expanded over the world and produced around 100,000 species in this short period of geological time. They are responsible for huge silica deposits, including diatomaceous earth, and have already played a substantial role in global carbon and nitrogen cycles (Gross, 2012).

5.2 DIATOMS SPECIES FOR CARBON SEQUESTRATION

5.2.1 Introduction to Diatoms Species

Diatoms are the most diverse group of algae, with more than 100,000 species. Diatoms are involved in approximately 20% of global carbon fixation and have a major influence on marine ecosystem primary production (Alindonosi et al., 2021; Falciatore et al., 2020). Diatoms are common in the plankton and benthos of freshwater and marine environments (Alindonosi et al., 2021). They can be described as unicellular algae with a set of genes and metabolic pathways. These diatoms have their physiological function obtained through endosymbiosis, which gives them the ability to be a photosynthetic carbon fixer (Liu et al., 2020).

In general, diatoms can be regarded as the most recent major algal lineage in addition to the existing algal lineage in the geological record that appeared since more than 200 million years ago. Diatoms are stramenopile protists that resulted from a secondary endosymbiotic event that produced the only photosynthetic protistan lineage with the characteristic of external siliceous cell wall structures known as frustules. Many diatoms are known to have large internal vacuoles that lead to the assumption of the diatoms' success for their capability for carbon fixation and survival are largely attributed to their two morphological characteristics: 1) defense provided by frustule and 2) luxury of nutrient uptake due to vacuole (Behrenfeld et al., 2021).

Diatom species can thrive in different environments, from polar to tropical waters. They evolved to adapt to different light environments and regulate light capturing and utilization to achieve successful dominance in the ecosystem (Fisher et al., 2020). Traditionally, the diatom species were studied by determining taxonomy, cellular composition, and functional roles that can be achieved through microscopic examination and cell culture technique. More recent identification and characterization of diatoms are done using molecular approaches such as metagenomics (Alindonosi et al., 2021). It is now understood that there are up to 25,000 diatom species that have been identified, and many more are yet to be discovered. It is quite challenging to identify diatom species and to classify them despite how experienced and skilled a person might be. There are guides to assist in diatom classification based on shape, morphology, and others (Williams, 2020).

Diatom species can be divided into two classifications based on their resilient siliceous cell walls (i.e., frustules) that consist of two valves based on their symmetry: 1) centric diatoms with radial symmetry and 2) pennate diatoms with bilateral symmetry (Robertson, 2013; Villanova & Spetea, 2021). A lot of studies have been conducted on various species of diatoms to understand their species classification and characteristics. Various taxon of diatoms were used in research to fully understand the morphology of diatoms, such as to determine phylogenetic position of toxarium within centric diatom species such as *Hyalosira sp. F. T. Kützinga, Grammatophora oceanica Ehrenberg, Grammatophora gibberula Kützing, Licmophora juergensii Agardh, Fragilaria crotonensis Kitton,* and *Thalassionema nitzschioide* (Kooistra et al., 2003).

Various species can be noted from research that include studies on 189 diatoms, including *Odontella sinensis, Papiliocellulus elegans, Cymatosira belgica, Biddulphiopsis titiana, Lampriscus kittonii, Toxarium undulatum, Eucampia antarctica, Chaetoceros rostratus, Asterionellopsis glacialis, Phaeodactylum tricornutum Pseudogomphonema sp., Thalassionema nitzschioides, Pleurosigma sp.,* and others, to understand the evolution of the elongated shape in diatoms (Alverson et al., 2006). More and more species of diatoms are being studied, not only for their diversity in the environment and their role in carbon fixation but also for their potential in industrial applications, generation of value-added products, and carbon footprint reduction.

5.2.2 Diatom Species Distribution in the Ecosystem

In the last several decades, international initiatives such as the European Marine Biological Resource Centre, Marine Genomics Europe, and the Marine Microbiology Initiative have been promoting research on diatom genetics and genomics. This has led to advancements in molecular model systems among phytoplankton organisms, for which previously diatoms could be regarded as having unknown cellular and metabolic processes. Molecular insights of diatoms generated from various studies helped to establish an emerging model for an identification and reference system of diatom species (Falciatore et al., 2020).

In general, diatoms have variation in regard of shapes and sizes, which may vary from the tiny Minidiscus spp. (about 3 μm in diameter) to some others that are almost visible to the naked eye. These diatoms have features known as frustules, valves, and girdle bands. In general, diatom valves are simple but may come in different varieties of structures. F dor example, diatom genera such as *Pleurosigma* and *Gyrosigma* are gently sigmoid, and fossil diatoms like *Kittonia elaborata* have tubular spines (Tiffany & Nagy, 2019).

Diatoms widely vary in size, morphology, and composition of the elements that control the quantity, quality, and sinking speed of carbon matter to depth. Particularly, their silica shells contribute to ballast to faecal pellets and marine snow and assist in transport of carbon to different layers of the ocean. This helps to sustain the marine food web through carbon export from the ocean surface to the deep ocean (Tréguer et al., 2018). Diatom taxonomy is determined based on the frustule morphology, and many species are originally described based on light microscopy observation (Schoefs et al., 2020). Examples of centric diatoms include *Skeletonema costatum, Cyclotella cryptica,* and *Thalassiosira pseudonana,* whereas examples of pennate diatoms are *Navicula saprophila, Nitzschia sp., Phaeodactylum Tricornutum,* and *Fistulifera solaris* (Falciatore et al., 2020; Villanova & Spetea, 2021).

Alverson et al. (2006) mentioned that diatoms have been historically classified as centric and pennate based on several features. However, the problem with this classification method is diatoms were highlighted by some recent analysis in regard to the Toxarium phylogenetic position of some diatoms, which concluded that some centric diatoms independently evolved with several pennate-like characteristics such as elongated cell shape. Regardless, there are thousands of species of diatoms, and using the morphology of the diatoms for species classification is still in place.

Phaeodactylum tricornutum is widely known to be distributed in the world. It is considered to be a model diatom species and is widely studied for its ability in respiration, photosynthesis, and other physiological performance (Tong et al., 2021). Besides its ability in carbon fixation in the environment, more study on this diatom is being conducted under the biotechnology field to determine its ability as a cell factory for various bio-based products such as eicosapentaenoic acid, neutral lipids, fucoxanthin, and crysolaminarin to establish an environmentally sustainable solution of a value-added product (Butler et al., 2020).

Diatoms rely on environmental variables to play their role in regulating photosynthetic carbon fixation in the ecosystem. Environmental variables may be indicated as external driving forces of diatoms for carbon sequestration. Among the ways to determine gross primary productivity of diatoms is through light and dark biological oxygen demands. Carbon sequestration potential was the highest in autumn. This was due to diatom production in the study area and the correlation between diatoms and phosphate in the environment (Yang et al., 2020).

Various diatom species have been studied to determine their photosynthetic strategy divergence in the environment. This includes three diatoms' taxa, namely *Thalassioisira weissflogii, Thalassiosira pseudonana,* and *Thalassiosira oceanica,* to understand their divergent strategies under various light environments (Fisher et al., 2020), which ensure their capability to produce oxygen and fix carbon in the ecosystem.

Moreover, diatoms have been known to survive polluted environments as well as various extreme environments, which includes highly acidic environments. According to Denicola (2000), in a review of 19 studies related to highly acidic environments with pH <3.5, a total of 124 diatom taxa can be detected in highly acidic environments, but from these, only approximately 19 taxa were found to be abundant in at least one study or common to be found in several studies as inhabitants of acidic water. These include the following species: *Nitzschia pusilla, Nitzschia communis, Nitzschia vasta, Eunotia tenella, E. exigua, E. osoresanensis, E. septentrionalis, E. glacialis, E. arcus, E. pectinalis, Achnanthes minutissima, Frustulia rhomboides, F. rhomboides var. saxonica, Nitzschia capitellata/subcapitellata, Pinnularia acoricola, P. braunii var. amphicephala, P. obscura, P. terminitina,* and *P. subcapitata.*

Diatom species may vary from one place to another depending to geographical and hydrological characteristics of an area, seasonal changes, and environmental factors. According to Ryabushko et al. (2019), species diversity of epilithon diatoms in a Crimean coastal water off the Black Sea found a total of 60 Bacillariophyta taxa, which can be categorized into 3 classes, 18 families, and 34 genera. From this finding, 45 species of diatoms were observed during summer, whereas a smaller number of species were found in the winter. Mass species found are pennate diatoms such as from the genera *Licmophora, Cylindrotheca, Grammatophora, Trachyneis, Striatella,* and *Cocconeis.*

A study on the Malili Lakes in Indonesia detected the diatom species of *Gomphonema* such as *G. longissimum, G. angustissimum, G. woltereckii,* and *G. subtiliforme* where these taxa are similar to each other and resemble other species of the genus. Despite the findings on this taxa from the ancient lake system, in actuality less than two dozen *Gomphonema* species are found in the Malili Lakes compared to other ancient lakes like Lake Baikal, where more species and flocks of diatoms can be identified (Kulikovskiy et al., 2019).

Diatoms' role in carbon fixation helps to sustain the marine food web in which they contribute to about 40% of marine primary productivity as well as act in carbon export from the surface of the ocean to the depth as part of the biological pump. Different species of diatoms can be found in abundance in water, especially those having turbulent and high nutrient contents (Tréguer et al., 2018). Table 5.1 provides the lineage of diatoms in relation to their net primary production and carbon export in the oligotrophic low-latitude ocean. Nonetheless, this table does not indicate all the species of diatoms that can be found in the ecosystem.

As previously mentioned, diatom species may vary from region to region, from area to area, and from river to ocean because diatom diversity can be influenced by geographical and hydrological characteristics of an area, together with their environmental factors and conditions. Table 5.2 lists the species-rich diatom taxa present in Cascadia Bioregion.

Benthic diatoms, which can be found and isolated from the Baltic Sea, may include species from taxa *Melosira moniliformis*, *Navicula perminuta*, *Halamphora sp.* 1, *Halamphora sp.* 2, *Actinocyclus octonarius*, *Navicula phyllepta*, *Nitzschia pusilla, and Nitzschia dubiiformis* (Prelle et al., 2019), whereas in Dajiuhu Peatland of central China, diatom communities were dominated by *Pinnularia* and *Eunotia* species from a total of 92 diatoms taxa identified from the area. These diatom communities' presence and growth have been linked with precipitation rate, water table depth, nitrates, conductivity, and temperature in the diatom's habitat (Chen et al., 2021).

In the Lalin River Basin, which is a major drainage basin located in northeastern China, a total of 50 diatom taxa were identified that belong to 25 genera of diatoms. Most of the identified taxa were pennate diatoms, and the dominant species found includes *Achnanthidium minutissimum, Cymbella tumida, Cocconeis placentula, Fragilaria vaucheriae var. capitellata, Encyonema silesiacum, Encyonema minutum, Gomphonema parvulum, Gomphonema acuminatum, Melosira varians, Hannaea arcus, Navicula viridula, Navicula viridula var. rostellata, Synedra ulna, Surirella angusta, Surirella visurgis,* and *Surirella minuta.* Most taxa identified are from *Navicula, Gomphonema, Fragilaria, Surirella, Nitzschia,* and *Cymbella* (Lu et al., 2020). The presence and

TABLE 5.1
Diatom Lineages Contributing to Diatom Diversity on the Ocean Biological Carbon Pump

Significantly Contributed to Net Primary Production and Carbon Export	Significantly Contributed to Carbon Export
Haslea spicula	*Grammatophora*
Thalassiosira sp.	*Stephanodiscus*
Pseudo-nitzschia multiseries	*Opephora guenter-grassii*
Thalassiosira weissflogii	*Bacillariophyta environmental lineages*
Pseudo-nitzschia	*Planktoniella sol*
Raphid pennate	*Polar centric Mediophyceae*
Pseudo-nitzschia heimii	*Thalassiosira Actinocyclus curvatulus*
Fragilaria crotonensis Pleurosigma	*Thalassiosira punctigera*
Radial centric *Coscinodiscophyceae Rhizosolenia Pseudo-nitzschia fraudulenta Araphid pennate Pseudo-nitzschia pseudodelicatissima Lithodesmium undulatum Radial centric Coscinodiscophyceae environm. lineage Craticula cuspidata Eucampia Guinardia flaccida Rhizosolenia shrubsolei Leptocylindrus*	
Radial centric Coscinodiscophyceae environm. lineage Chaetoceros rostratus	
Synedra	

Other diatoms species that influence the ocean biological carbon pump

Synedra toxoneides	*Bacillariophyta environm. Lineage*
Cylindrotheca clostrerium Bacillariophyta environm. lineage	*Lauderia borealis*
Navicula perminuta	*Chaetoceros radicans*
Thalassiosira hendeyi	*Pseudo-nitzschia delicatissima*
Cymbella	*Skeletonema grevillei Corethron hystrix*
Cocconeis stauroneiformis	*Pseudo-nitzschia australis*
Raphid pennate environm. lineage	*Leptocylindrus convexus*
Thalassiosira minima	*Navicula radiosa Asterionellopsis glacialis Raphoneis sp.*
Thalassiosira rotula	*Skeletonema menzellii*
Nitzschia apiculata	*Coscinodiscus sp.*
Chaetoceros sp.	*Skeletonema*
Cyclotella choctawhatcheeana Actinoptychus sinensis	*Minidiscus*
Coscinodiscus radiatus	*Odontella sinensis*
Thalassiosira delicatula	*Naviculales sp.*
Corethron	*Stellarima microtrias*
Sellaphora seminulum	*Synedra sp.*
Thalassiosira aestivalis	*Polar centric Mediophyceae environm. lineage*
Hyalosira sp.	*Minutocellus*
Thalassiosira concaviuscula	*Ditylum brightwellii*
Pseudo-nitzschia sp.	*Pseudohimantidium pacificum*
Chaetoceros	*Proboscia alata*

Source: (Tréguer et al., 2018).

density of diatoms in different regions and locations can be influenced by the presence of nutrients in the water, seasonal patterns, precipitation, geographical and hydrological characteristics, and environmental conditions that can be proven as challenges to understanding the ability of diatom species in carbon fixation.

TABLE 5.2
Species-rich Diatom Taxa Present in Cascadia Bioregion

Taxa	Species
Encyonema species with large cells and coarse striae (*Encyonema muelleri sensu lato*)	*Encyonema kamtschaticum var. kamtschaticum Krammer*
	Encyonema kamtschaticum var. parva Krammer
	Encyonema latum Krammer
	Encyonema montana Bahls
	Encyonema nicafei Spaulding
	Encyonema reimeri Spaulding, Pool & Castro
	Encyonema temperei Krammer
	Encyonema yellowstonianum Krammer
Gomphoneis Cleve (G. herculeana and *G. septa Groups)*	*Gomphoneis compacta Bahls*
	Gomphoneis eriensis var. eriense (Grunow) Skvortzow & Meyer
	Gomphoneis eriense var. angularis Kociolek & Stoermer
	Gomphoneis eriense var. variabilis Kociolek & Stoermer
	Gomphoneis herculeana var. herculeana(Ehrenberg) Cleve
	Gomphoneis herculeana var. abundans Kociolek & Stoermer
	Gomphoneis herculeana var. lowei Kociolek & Stoermer
	Gomphoneis herculeana var. minuta Stone
	Gomphoneis linearis Kociolek & Stoermer
	Gomphoneis mammilla (Ehrenberg) Cleve
	Gomphoneis montana Bahls
	Gomphoneis 'salishorum' (provisional name)
	Gomphoneis septa (Moghadam) Kociolek et al
Gomphonema species with wide axial areas *(G. rhombicum Fricke* in Schmidt et al. *sensu lato)*	*Gomphonema amerhombicum Reichardt*
	Gomphonema 'california' (provisional name)
	Gomphonema caperatum Ponader & Potapova
	Gomphonema incognitum Reichardt, Jüttner & Cox
	Gomphonema sierrianum Stancheva & Kociolek
	Gomphonema stoermeri Kociolek & Kingston
Kurtkrammeria Bahls	*Kurtkrammeria aequaliformis (Bahls) Bahls*
	Kurtkrammeria aequalis (W. Smith) Bahls
	Kurtkrammeria lacusglacialis (Bahls) Bahls
	Kurtkrammeria neoamphioxys (Krammer) Bahls
	Kurtkrammeria stodderi (Cleve) Bahls
	Kurtkrammeria subspicula (Krammer) Bahls
	Kurtkrammeria treinishii (Bahls) Bahls
	Kurtkrammeria weilandii (Bahls) Bahls

Source: Bahls (2019).

5.3 CARBON SEQUESTRATION BY DIATOMS

5.3.1 Introduction to Carbon Sequestration

Diatoms are *bacillariophytes* with silicified cell walls that are one of the most diverse classes of photosynthetic primary producers in the coastal and marine environment, with 30,000 to 100,000 species found in both freshwater and marine environments. Diatoms can constitute a large amount in sediment, which can be accurately quantified using quantitative counts of valve numbers and conversion of abundant values to biovolume and accumulation rates if biogenic silica is adequately maintained. They are commonly encountered as microalgae because of their ability to live in a variety of physicochemical conditions, such as growing under sea ice or in thermal springs. They are reported to store carbon as chrysolaminarin in the vacuole and have a pigment profile that includes chlorophyll and fucoxanthin as the major light harvesting complex, rather than chlorophyll as in other plants. Diatoms are considered to provide at least 20% of global CO_2 uptake and play a significant role in carbon, nitrogen, and silicon biogeochemical cycling. In addition, manipulation of carbon sequestration by diatom can also be used to improve the accumulation of both lipids and carbohydrates, as has been studied in *T. pseudonana, P. tricornutum, Asterionella formosa,* and *Navicula pelliculosa.* One of the driving factors of diatom metabolic versatility is evolutionary transitions to obtain energy from different trophic modes.

5.3.2 Carbon Biosequestration by Diatoms

The intervals on which CO_2 is taken up by phytoplankton is sequestered from the atmosphere depending on the depths at which microorganisms and zooplankton remineralize organic materials sinking out of the surface layer back to CO_2. Phytoplankton are essential contributors to organic carbon production because of their rapid growth rates in high-nutrient environments and comparatively large proportions. Diatoms play an important role in marine food chains and in the sequestration of atmospheric CO_2 into the ocean interior via gravitational sinking of particles, making them biogeochemically valuable. Generally, the ribulose-1,5-bisphosphate carboxylase/oxygenase, Rubisco, the major carbon assimilation enzyme, is one of the most prevalent proteins on Earth and is responsible for 100 billion tonnes of carbon assimilation per year. Remarkably, however, diatoms that consist of a red algal type Rubisco (type ID) are one of the most effective Rubisco forms, favoring the processes of carboxylation the most. Previous experimental studies show that diatoms can occupy both CO_2 and HCO_3 and precipitate dissolved inorganic carbon in their cells. Wu et al. (2014) recently discovered that higher CO_2 stimulates the growth of larger vs. smaller centric diatoms in distinct ways, presumably due to the removal of diffusional constraints on CO_2 transport to the surfaces of larger cells. This shows that the cytosolic metabolic pathways in diatoms are more closely related to an animal-like cell than to a plant-like cell, which would have an impact on photosynthetic physiological modulation. Diatoms, a prominent phytoplankton group, are thought to provide about 40% of particulate organic carbon export and have a substantial impact on the biological carbon pump's productivity. Diatoms could aid with organic carbon assimilation in a number of ways, including their considerable contribution to oceanic primary production, their proliferation propensity and subsequent marine ranching, their vast sizes, and the critical role of silica frustules in protective function, as well as ballasting sinking particles. Table 5.3 below shows the performance of diatoms in sequestration CO_2 from the atmosphere. Referring to Table 5.3, *Skeletonema* sp. and *Chaetoceros* sp. have exhibited good performance in their growth at varied CO_2 concentrations in the ocean, whether in still or turbulent conditions. *Chaetoceros* sp. has shown better CO_2 sequestration if compared to *Skeletonema* sp., even though the cell abundance is lesser for the former. *Thalassiosira pseudonana, Thalassiosira oceanica,* and *Thalassiosira weissflogii* have shown CO_2 removal of up to about 30% at lab scale experiment, with the condition of alumnium-enriched treatment. More experimental cultivation conditions are required to further investigate their potential in removing CO_2. Further, mixture or consortium of diatoms is shown to exhibit outstanding CO_2 biosequestration, with up to 70% CO_2 removal rate. Considerably good cell division performance was also being observed. Overall, we could have observed the performance and potential of diatoms in biosequestrating CO_2. Nevertheless, it should also be noted that there is no strict comparison among the diatoms documented, due to the variation in experimental conditions.

TABLE 5.3
The Performance of CO_2 Biosequestration by Diatoms

No	Diatom	Condition	Initial CO_2 Concentration (%)(v/v)	Removal Percentage (%)	CO_2 Bio-fixation Rate ($g\ L^{-1}\ d^{-1}$)/($g\ m^{-2}\ d^{-1}$)^a/($g\ CO_2\ d^{-1}$)^b	Abundance ($Cell\ L^{-1}$)/Biomass Productivity ($g\ L^{-1}\ d^{-1}$)^a/Biomass Productivity ($g\ m^{-2}\ d^{-1}$)^b/Biomass Yield ($g\ L^{-1}$)^c/Cell (Division d^{-1})^d	References
1	*Chaetoceros*	Ocean, still conditions	Atmospheric	10	–	1.486×10^6	(Bergkvist et al., 2018)
2	*Chaetoceros*	Ocean, turbulent shear	Atmospheric	15	–	1.486×10^6	(Bergkvist et al., 2018)
3	*Phaeodactylum tricornutum*	Batch operation	15	-	0.28	0.15^a	(Sethi et al., 2020)
4	*Phaeodactylum tricornutum*	Photobioreactor	–	-	35.5^a	21 ± 2.3^b	(Sethi et al., 2020)
5	*Phaeodactylum tricornutum*	Lab scale flask method	1.8–2.4	50–65	$0.06–0.08^c$	$\sim 1.0^c$	(Piiparinen et al., 2018)
6	*Skeletonema*	Ocean, still conditions	Atmospheric	2	–	2.023×10^6	(Bergkvist et al., 2018)
7	*Skeletonema*	Ocean, turbulent shear	Atmospheric	4	–	2.023×10^6	(Bergkvist et al., 2018)
8	*Thalassiosira pseudonana*	Lab scale flask method	1.8–2.4	50–65	$0.06–0.08^b$	$\sim 1.0^c$	(Piiparinen et al., 2018)
9	*Thalassiosira pseudonana*	Lab scale method, aluminum enriched treatment	–	9–29	–	–	(Zhou et al., 2021)
10	*Thalassiosira oceanica*	Lab scale method, aluminum enriched treatment	–	15–20	–	–	(Zhou et al., 2021)
11	*Thalassiosira weissflogii*	Lab scale method, aluminum enriched treatment	–	15–23	–	–	(Zhou et al., 2021)
12	*Prymnesiophytes*, *prasinophytes*, and small diatoms	Ocean	Atmospheric	41–70	–	0.37 ± 0.13^d 0.22 ± 0.09^d	(Irion et al., 2021)

5.4 PHOTO-PHYSIOLOGICAL RESPONSES OF DIATOMS

When combined with bioprocessing and biomanufacturing for added value, CO_2 sequestration by photosynthetic organisms can also be a reasonable solution. Due to its superior photophysiology, diatoms are preferred over other phytoplankton groups in environments with fluctuating light, such as non-stratified water columns, as demonstrated by *Phaeodactylum tricornutum* and *Thalassiosira weissflogii*. However, environmental temperature is critical for almost all marine organisms, as the fitness of marine organisms is expected to be impacted because their metabolism and cellular, molecular, or enzyme activities are all temperature dependent, reflecting their physiological performance. As macroalgae from different habitats are adapted to different temperature ranges, the response to global warming is species specific. For instance, the growth of *Pyropia haitanensis* was not stimulated by rising growth temperatures from 18°C to 22°C. In a low temperature region, biologically harmful levels of solar ultraviolet radiation (UVR) may penetrate through the perennial ice cover and shallow water column because of the relatively low concentration of UVR absorbing materials. Such conditions, combined with strong and continuous solar radiation during the summer, may affect the growth of diatom phototrophs living in the shallow parts of the lakes through inhibition and photo-physiological damage. Meanwhile, in a higher temperature region, sustained nutrient depletion will potentially reduce photosynthetic phytoplankton biomass and marine primary productivity, and a shallower UML increases solar radiation for the organisms living in it. Stratification can also be enhanced in inner bays during summer periods, and in the coastal system, when repeated onshore circulation of water is unfavorable for external sources of nutrients during summer, nutrient availability will be limited for the growth of diatom species.

5.5 OCEAN ACIDIFICATION RESPONSE ON DIATOMS

Our oceans have absorbed approximately one-third of all anthropogenic CO_2 released into the atmosphere, resulting in altered marine carbonate chemistry with a continuous decline of pH, a process known ocean acidification (OA). OA that is caused by continuous dissolution of anthropogenic CO_2 emissions into the oceans, has been suggested to impact marine diatom primary productivity. There are numerous works showing that elevated CO_2 concentration can stimulate diatom growth and carbon fixation. For example, the effect of OA on the growth of the diatom *Thalassiosira pseudonana* was sometimes negligible or interacted with growth light. Many studies have observed a stimulation, inhibition, or neutral effect of OA on growth or primary production in diatom-dominated communities under OA, leading to controversy worth further investigation. Projections suggest that by the year 2100, average atmospheric CO_2 will increase between 720–1,020 ppm, with the resultant decrease in surface ocean pH by 0.14–0.32 pH units. The potential impacts of OA are likely magnified in these eastern boundary upwelling regions due to the key role these regions play in supporting the bulk of the world's fisheries. Declining pH can affect intracellular acid-base balance and nutrient absorption as well as reduce algae growth under nutrient-limited conditions or under high levels of sunlight. pH drops decrease mineralization of diatoms and coccolithophores. On the other hand, increased availability of HCO_3 though intracellular dissolved inorganic carbon concentration can actually decrease under climate change relevant CO_2 levels when down-regulation of CCMs outweighs increased passive influx of DIC into cells.

5.6 CHALLENGES AND OPPORTUNITIES OF DIATOMS IN CARBON FOOTPRINT REDUCTION

Diatoms can be included in the intertidal microbial communities, which can present as biofilms or microphytobenthos (MPB) that are attaching to sediment assemblages containing bacteria, protozoa, fungi, algae, and diatoms. Diatoms, together with these microbial communities, provide global annual productivity of 5,000 million tons of carbon in which 50% of this carbon fixation contributes toward estuaries (Mandal et al., 2021). This indicates the importance of these microbial communities, including diatoms, in carbon dioxide sequestration not only for acting as natural sink for carbon dioxide for footprint reduction but also for their potential in heavy metal removal.

There is urgent need to reduce the carbon footprint in the environment due to the warming of the earth's surface, leading to various environmental consequences. It is well-known that CO_2 is one of the major greenhouse gases notorious for causing climate change. Rising interest has been placed on diatoms, which have been known as natural atmospheric carbon sinks. More and more studies were conducted to cultivate diatoms industrially in either autotrophic or mixotrophic modes for CO_2 sequestration. Besides, diatoms exhibit metabolic diversity in providing ways to generate valuable products from captured carbon. Among targeted products of carbon sequestration by diatoms that can be produced under environmentally friendly and sustainable biomanufacturing processes are exopolysaccharides, lipids, antioxidants, omega-3 fatty acids, sulphated polysaccharides, pigments, and other metabolites (Sethi et al., 2020).

Diatoms can survive even in polluted water and have the capability for pollutant bio-absorption. Aside from their role in carbon footprint reduction in the world, diatoms can be utilized for pollutant remediation in the environment. This has caused growing interest in diatoms for toxic metal bioremediation. Reasons for the increasing interest in diatoms for bioremediation are that 1) diatoms are one of the most abundant and dominant groups of phytoplankton in oceans, 2) they are species that are tolerant to pollution, 3) they act as excellent water quality biological indicators, and 4) they have efficient models in toxic metal ion assimilation and detoxification (Chasapis et al., 2022).

According to Chen et al. (2021), changes in diatoms and chemical elements in the environment can be closely associated with rising temperatures and other environmental factors such as variability in precipitation, atmospheric deposition, and local hydrogeomorphic setting. This can pose a potential increase and decrease in terms of diatom communities, hence influencing carbon fixation and primary production of affected areas. It was mentioned that alarming future warming of the climate and heavy rainfall variation are expected to alter and influence diatom communities and their biogeochemical processes in the subtropical peatland ecosystems.

Studies has suggested that zinc, which is an essential micronutrient for algae growth, including diatoms, may influence the growth of diatoms due to scarce concentrations of dissolved zinc in the ocean surface. This may affect diatoms' growth, thus affecting primary productivity in the ocean as diatoms and phytoplankton are known to be the main contributors to ocean primary productivity. In the case of zinc scarcity, specific adaptive mechanisms of diatoms to this deficiency remain poorly understood (Kellogg et al., 2022). This calls for more studies to understand multiple adaptive responses of diatoms to zinc scarcity in marine ecosystems, for example, in low-zinc regions in certain oceans.

Diatoms can alter marine food webs due to the increase of carbon dioxide in the environment, particularly in marine ecosystems. According to Harvey et al. (2019), seabed habitats were seen to be significantly changed due to elevated concentrations of CO_2 in the ocean. This increasing CO_2 stimulates the growth and photosphysiology of certain species of diatoms such as *Biddulphia biddulphiana,* causing it to become a dominant benthic habitat species. This ecological shift may increase the rate of photosynthesis but lead to algal bloom that has the capability to support the abundant fauna community. This leads to ecological shifts in the marine ecosystem, which may impact the food web structure and general ecosystem functioning.

Diatom species may become dominant in areas with elevated CO_2 levels, whereas macroalgae became dominant at current present-day CO_2 levels. This can be proven by diatoms' pigment composition, fucoxanthin content, and characteristics, which seem to be higher at the high CO_2 site. Algal communities, including diatoms found in water with high levels of CO_2, can have elevated rates of photosynthesis. Major carbon fixation by algae in areas with high concentration of CO_2 was exported due to detachment from the different level of ocean depth. The reason is because diatoms that dominated under this condition are more easily being transported away as these diatoms have no holdfast to elevated CO_2 concentration. This shows that ocean acidification may alter coastal carbon cycling, increasing photosynthesis and carbon export from coastal ecosystems in warm-temperate biogeographic regions due to a shift in community composition from perennial to ephemeral algae, including diatoms. Such changes may reduce algal communities' habitat complexity and their biomass in rocky shore habitats unless human beings can provide solutions to minimize and address the impact of ocean acidification (Wada et al., 2021). This alteration of algae communities, including diatoms, may cause major disruptions to the coastal ecosystem, including toward food webs, fisheries, and the carbon cycle.

However, under current CO_2 conditions, it is expected that the biological pump efficiency that is currently mediating carbon sequestration in the environment will reduce and decline in the future. The reason is that respiration of diatoms tends to be more sensitive to rising temperatures than is photosynthesis. Nonetheless, the impact of global warming on the metabolic rates of diatoms still remains unclear and may change due to the elevated CO_2 induced by ocean acidification, but findings from the research indicate that ocean acidification may worsen the negative impact and have the capability to offset the alleviating impact of warming respiration to photosynthesis ratio. This depends on the temperature range of diatoms such as *Phaeodactylum tricornutum* (Tong et al., 2021).

In the reliance on diatoms as carbon sinks, these microalgae can be used as potential feedstocks for production of high-value lipids, carotenoids, and biofuels. Despite requiring more study to fully understand the main metabolic pathways for diatoms, especially those exhibiting mixotrophy properties, these diatoms have potential in industrial applications due to their ability to produce triacyl glycerides, poly-unsaturated fatty acid, and other molecules of interest, and the cost of organic supplements can be obtained through the use of wastewater and biodiesel waste as carbon supplements in industrial applications (Villanova & Spetea, 2021).

Further understanding of environmental effects and conditions can provide more insight on the role of diatoms in carbon footprint reduction. Environmental condition may include temperature level, light availability, and precipitation rate in an area (Chen et al., 2021; Prelle et al., 2019). Studies conducted on benthic diatoms indicate that diatom species such as *Actinocyclus octonarius, Melosira moniliformis, Navicula perminuta,* and *Nitzschia pusilla* have high tolerance against fluctuating environmental conditions in the shallow coastal area of the Baltic Sea (Prelle et al., 2019). This finding is important to understand the implication of diversity of diatoms and their tolerance to environmental changes, which may affect the primary production rates and photosynthesis of diatoms.

Throughout this chapter, it has been shown that diatoms are successful microalgae (at least one of them) contributing to global primary productivity and playing a role in marine and freshwater nutrient cycles. Diatoms have the ability to biosynthesize value-added substances such as eicosatetraenoic acid (EPA), fucoxanthin (Fx), and chrysolaminarin (Chrl). Diatoms have a unique potential in obtaining diverse carbon sources via various mechanisms, which gives them the ability to act as phototrophic, mixotrophic, and heterotrophic modes in different conditions. Diatom growth utilizing organic carbon can promote higher production of biomass, lipid, and carbohydrate, which triggers the generation and yield of various biomolecules (Marella et al., 2021). This has become the focus of diatom research, in addition to the ability of diatoms to aid in carbon fixation, which helps in the reduction of the carbon footprint in the atmosphere.

Nonetheless, diatoms can be utilized to target climate recovery. This can happen through carbon dioxide reduction strategies, carbon capture and storage, carbon capture utilization and storage, and carbon capture and utilization conducted by utilizing the microalgae. In general, microalgal cells that include diatoms will utilize CO_2 as precursors of various macromolecules (e.g., proteins, lipids, and carbohydrates) that have the potential for commercialization and industrial application. The use of microalgae for carbon fixation can serve as pathways toward carbon footprint reduction, and it can be achieved via carbon capture before carbon being emitted to the atmosphere. Increases in the carbon handprint in which carbon emission reduction through carbon dioxide neutral or negative products consumption may help in carbon footprint reduction (Daneshvar et al., 2022).

5.7 CONCLUSION

The excessive emission of CO_2 from anthropogenic activities has created a threat to the human race and ecosystem. Incidents of floods, landslides, forest fires, etc., have occurred more frequently. Environmental sustainability is the main goal for the human race to achieve in this era. CO_2 sequestration via biological resources, aside from terrestrial plants, would be algae species, including diatoms. Exploration of the diatom species, environmental conditions for their successful cultivation, as well as diatom biorefineries for industrial applications needs further advancement. This aims toward the environmental sustainability and well-being of society.

REFERENCES

Alindonosi, A. R., Baeshen, M. N., & Elsharawy, N. T. (2021). Prospects for diatoms identification using metagenomics: A review. *Applied Ecology and Environmental Research*, 19(6), 4281–4298. 10.15666/aeer/1906_42814298

Alverson, A. J., Cannone, J. J., Gutell, R. R., & Theriot, E. C. (2006). The evolution of elongate shape in diatoms. *Journal of Phycology*, 42(3), 655–668. 10.1111/j.1529-8817.2006.00228.x

Babacan, O. et al. (2020). Assessing the feasibility of carbon dioxide mitigation options in terms of energy usage. *Nature Energy*, 5(9), 720–728. 10.1038/s41560-020-0646-1.

Bach, L. T. et al. (2019). Effects of elevated CO_2 on a natural diatom community in the subtropical NE Atlantic. *Frontiers in Marine Science*, 6(MAR), 1–16.

Bahls, L. L. (2019). Diatoms from Western North America 3. Signature Diatoms of the Cascadia Bioregion Address of the author: (Issue September).

Behrenfeld, M. J., Halsey, K. H., Boss, E., Karp-Boss, L., Milligan, A. J., & Peers, G. (2021). Thoughts on the evolution and ecological niche of diatoms. *Ecological Monographs*, 91(3). 10.1002/ecm.1457

Bergkvist, J., Klawonn, I., Whitehouse, M. J., Lavik, G., Brüchert, V., & Ploug, H. (2018). Turbulence simultaneously stimulates small-and large-scale CO_2 sequestration by chain-forming diatoms in the sea. *Nature communications*, 9(1), 1–10. 10.1038/s41467-018-05149-w

Butler, T., Kapoore, R. V., & Vaidyanathan, S. (2020). *Phaeodactylum tricornutum*: A diatom cell factory. *Trends in Biotechnology*, 1–17. 10.1016/j.tibtech.2019.12.023

Chasapis, C. T., Peana, M., & Bekiari, V. (2022). Structural identification of metalloproteomes in marine diatoms, an efficient algae model in toxic metals bioremediation. *Molecules*, 27(2), 1–10. 10.3390/molecules27020378

Chen, X., McGowan, S., Qin, B., Huang, X., Stevenson, M. A., Zhang, Z., Zeng, L., & Bai, X. (2021). Effects of climate change on a subtropical montane peatland over the last two centuries: Evidence from diatom records. *Holocene*, 31(7), 1112–1123. 10.1177/09596836211003220

Daneshvar, E., Wicker, R. J., Show, P., & Bhatnagar, A. (2022). Biologically-mediated carbon capture and utilization by microalgae towards sustainable CO_2 biofixation and biomass valorization – A review. *Chemical Engineering Journal*, 427(May 2021), 130884. 10.1016/j.cej.2021.130884

Denicola, D. M. (2000). A review of diatoms found in highly acidic environments. Hydrobiologia, 433, 111–122. 10.1023/A:1004066620172

Falciatore, A., Jaubert, M., Bouly, J. P., Bailleul, B., & Mock, T. (2020). Diatom molecular research comes of age: Model species for studying phytoplankton biology and diversity[open]. *Plant Cell*, 32(3), 547–572. 10.1105/tpc.19.00158

Fisher, N. L., Campbell, D. A., Hughes, D. J., Kuzhiumparambil, U., Halsey, K. H., Ralph, P. J., & Suggett, D. J. (2020). Divergence of photosynthetic strategies amongst marine diatoms. *PLoS ONE*, 15(12 December), 1–27. 10.1371/journal.pone.0244252

Gross, M. (2012). The Mysteries of the Diatoms. *Current Biology*, 22(15), R581–R585. 10.1016/j.cub.2012.07.041.

Harvey, B. P., Agostini, S., Kon, K., Wada, S., & Hall-spencer, J. M. (2019). Diatoms Dominate and Alter Marine Food-Webs When CO_2 Rises.Diversity,11(12),1-13.10.3390/d11120242

Irion, S., Christaki, U., Berthelot, H., L'helguen, S., & Jardillier, L. (2021). Small phytoplankton contribute greatly to CO_2-fixation after the diatom bloom in the Southern Ocean. *The ISME journal*, 15(9), 2509–2522. DOI: 10.1038/s41396-021-00915-z.

Jordan R. et al. (2020). Introduction. UCL Homepage: 1–10. https://www.ucl.ac.uk/GeolSci/micropal/diatom.html.

Kellogg, R. M., Moosburner, M. A., Cohen, N. R., Hawco, N. J., McIlvin, M. R., Moran, D. M., DiTullio, G. R., Subhas, A. V., Allen, A. E., & Saito, M. A. (2022). Adaptive responses of marine diatoms to zinc scarcity and ecological implications. *Nature Communications*, 13(1). 10.1038/s41467-022-29603-y

Kooistra, W. H. C. F., Stefano, M. De, Mann, D. G., Salma, N., & Medlin, L. K. (2003). Phylogenetic position of Toxarium, a pennate-like lineage within centric diatoms (BACILLARIOPHYCEAE). October 2017. 10.1046/j.1529-8817.2003.02083.x

Kulikovskiy, M., Kapustin, D., Glushchenko, A., Sidelev, S., Maltsev, Y., Gusev, E., Kezlya, E., Shkurina, N., Kociolek, P., Kulikovskiy, M., Kapustin, D., Glushchenko, A., & Sidelev, S. (2019). Morphological and molecular investigation of *Gomphonema longissimum* and related taxa from Malili lakes (Indonesia) with comments on diatom evolution in ancient lakes. *European Journal of Phycology*, 55(2), 147–161. 10.1080/09670262.2019.1664771

Liu, Q., Xing, Y., Li, Y., Wang, H., Mi, T., Zhen, Y., & Yu, Z. (2020). Carbon fixation gene expression in *Skeletonema marinoi* in nitrogen-, phosphate-, silicate-starvation, and low-temperature stress exposure. *Journal of Phycology*, 56(2). 10.1111/jpy.12936

Lu, X., Liu, Y., & Fan, Y. (2020). Diatom Taxonomic Composition as a Biological Indicator of the Ecological Health and Status of a River Basin under Agricultural Influence.Water,12(7),2067. 10.3390/w12072067

Mandal, A., Dutta, A., Das, R., & Mukherjee, J. (2021). Role of intertidal microbial communities in carbon dioxide sequestration and pollutant removal: A review. *Marine Pollution Bulletin*, 170(June), 112626. 10.1016/j.marpolbul.2021.112626

Marella, T. K., Bhattacharjya, R., & Tiwari, A. (2021). Impact of organic carbon acquisition on growth and functional biomolecule production in diatoms. *Microbial Cell Factories*, 1–13. 10.1186/s12934-021-01627-x

Medlin, L. K. (2016). Evolution of the diatoms: Major steps in their evolution and a review of the supporting molecular and morphological evidence. *Phycologia*, 55(1), 79–103.

Parry, I. (2015). Summary for policymakers. *Implementing a US Carbon Tax: Challenges and Debates*, 1(August): xxiii–xxxiii.

Piiparinen, J., Barth, D., Eriksen, N. T., Teir, S., Spilling, K., & Wiebe, M. G. (2018). Microalgal CO_2 capture at extreme pH values. *Algal research*, 32, 321–328. 10.1016/j.algal.2018.04.021

Prelle, L. R., Graiff, A., Gründling-Pfaff, S., Sommer, V., Kuriyama, K., & Karsten, U. (2019). Photosynthesis and respiration of baltic sea benthic diatoms to changing environmental conditions and growth responses of selected species as affected by an adjacent peatland (Hütelmoor). *Frontiers in Microbiology*, 10(JULY), 1–19. 10.3389/fmicb.2019.01500

Robertson, J. (2013). Botany. Encyclopedia of Forensic Sciences: Second Edition, 217–221. 10.1016/B978-0-12-382165-2.00115-X

Ryabushko, L. I., Lishaev, D. N., & Kovrigina, N. P. (2019). Species diversity of epilithon diatoms and the quality of the waters of the Donuzlav gulf ecosystem (Crimea, the Black sea). *Diversity*, 11(7). 10.3390/d11070114

Schoefs, B., Van de Vijver, B., Wetzel, C. E., & Ector, L. (2020). Introduction: From diatom species identification to ecological and biotechnological applications. *Botany Letters*, 167(1), 2–6. 10.1080/23818107.2020.1719883

Sethi, D., Butler, T. O., Shuhaili, F., & Vaidyanathan, S. (2020). Diatoms for carbon sequestration and bio-based manufacturing. *Biology*, 9(8), 1–29. 10.3390/biology9080217

Sethi, D., Thomas, O. B., Shuhaili, F., and Vaidyanathan, S. (2020). Diatoms for Carbon Sequestration and Bio-Based Manufacturing. *Biology*, 9(8), 1–29.

Tiffany, M. A., & Nagy, S. S. (2019). The Beauty of Diatoms. In S. Joseph & G. Richard (Eds.), *Diatoms: Fundamentals and Applications*. Wiley. 10.1002/9781119370741

Tong, S., Xu, D., Wang, Y., Zhang, X., Li, Y., Wu, H., & Ye, N. (2021). Influence of ocean acidification on thermal reaction norms of carbon metabolism in the marine diatom Phaeodactylum tricornutum. *Marine Environmental Research*, 164(December 2020), 105233. 10.1016/j.marenvres.2020.105233

Tréguer, P., Bowler, C., Moriceau, B., Dutkiewicz, S., Gehlen, M., Aumont, O., Bittner, L., Dugdale, R., Finkel, Z., Iudicone, D., Jahn, O., Guidi, L., Lasbleiz, M., Leblanc, K., Levy, M., & Pondaven, P. (2018). Influence of diatom diversity on the ocean biological carbon pump. *Nature Geoscience*, 11(1), 27–37. 10.1038/s41561-017-0028-x

Villanova, V., & Spetea, C. (2021). Mixotrophy in diatoms: Molecular mechanism and industrial potential. *Physiologia Plantarum*, 173(2), 603–611. 10.1111/ppl.13471

Wada, S., Agostini, S., Harvey, B. P., Omori, Y., & Hall-Spencer, J. M. (2021). Ocean acidification increases phytobenthic carbon fixation and export in a warm-temperate system. *Estuarine, Coastal and Shelf Science*, 250(May 2020), 107113. 10.1016/j.ecss.2020.107113

Williams, D. M. (2020). Diatom Classifications: What Purpose Do They Serve? In G. Cristóbal, B. Saúl, & G. Bueno (Eds.), *Modern Trends Identification in Diatom: Fundamentals and Applications* (pp. 11–24). Springer. 10.1007/978-3-030-39212-3

Wu, Y., Jeans, J., Suggett, D. J., Finkel, Z. V., and Campbell, D. A. (2014). Large centric diatoms allocate more cellular nitrogen to photosynthesis to counter slower RUBISCO turnover rates. *Frontiers in Marine Science*, 1(68), 1–11.10.3389/fmars.2014.00068.

Yang, J., Wang, F., Lv, J., Liu, Q., Nan, F., Liu, X., Xu, L., Xie, S., & Feng, J. (2020). The spatiotemporal contribution of the phytoplankton community and environmental variables to the carbon sequestration potential in an urban river. *Environmental Science and Pollution Research*, 27(5), 4814–4829. 10.1007/s11356-019-07109-3.

Zhou, L., Liu, F., Liu, Q., Fortin, C., Tan, Y., Huang, L., & Campbell, P. G. (2021). Aluminum increases net carbon fixation by marine diatoms and decreases their decomposition: Evidence for the iron–aluminum hypothesis. *Limnology and Oceanography*, 66(7), 2712–2727. 10.1002/lno.11784.

6 Aquatic Role of Diatoms

From Primary Producers and Aquafeeds

S.D. Ishak
HICoE, Institute of Tropical Aquaculture and Fisheries, Universiti Malaysia Terengganu, Terengganu, Malaysia

W. Cheah
Institute of Ocean and Earth Sciences, Universiti Malaya, Kuala Lumpur, Malaysia

K. Waiho
HICoE, Institute of Tropical Aquaculture and Fisheries, Universiti Malaysia Terengganu, Terengganu, Malaysia

S. Salleh
Center for Policy Research and International Studies, Universiti Sains Malaysia, Penang, Malaysia

H. Fazhan, H. Manan, N.A. Kasan, and S.S. Lam
HICoE, Institute of Tropical Aquaculture and Fisheries, Universiti Malaysia Terengganu, Terengganu, Malaysia

CONTENTS

DOI: 10.1201/9781003322115-6

6.1 INTRODUCTION

Microorganisms have been long exploited for their roles in nature. Single-cell microbes such as yeast and bacteria have been extensively researched for various industrial applications but they have their limitations. Diatoms, on the other hand, have the ability to adapt and grow in various bodies of water in any conditions, be it marine or freshwater, in the coldest regions on earth, or in the hottest deserts. Their ability to adapt in any aquatic environments makes them suitable to be exploited as a sustainable resource or raw material. The most distinctive appeal in using diatoms is that their growth can be easily influenced, the production period can be accelerated, and then they can be upscaled to meet demands. In this chapter, we discuss the natural roles of diatoms in the ecosystem and their aquatic role in benefitting mankind.

6.2 ROLES OF DIATOMS IN THE ECOSYSTEM

6.2.1 Pelagic and Benthic Diatoms

Diatoms are extremely diverse taxonomically. It has been estimated that there are between 10,000 and 100,000 species of living diatoms (Kooistra et al., 2007) existing in almost all climate and aquatic environments, ranging from the tropics to the Arctic and Antarctic. Vertically, diatoms are most commonly found floating in the sunlit layer of the upper ocean, and about 10,000 species have been identified (Malviya et al., 2016). Nonetheless diatoms can also be found as benthos in the sediments or trapped underneath the sea ice (McMinn et al., 2010); in this case, they are regarded as benthic microalgae instead of phytoplankton. The pelagic or planktonic diatoms usually triumph in waters rich in nutrients with sufficient sunlight, but recent studies record the presence of diatoms in the deep ocean down to 4,000 m (Agustí et al., 2015, Cedhagen et al., 2014). They can have a tremendous range of size, from < 1 μm to a few millimetres (Snoeijs et al., 2002), but the pelagic form is mostly confined to the size group of microphytoplankton with cell size between 20 and 200 μm.

Pelagic diatoms are the main actor of blooms in the open oceans such as the North Atlantic Bloom (Rynearson et al., 2013) and in the Southern Ocean (Deppeler and Davidson, 2017); however, these blooms are usually dominated by only a small fraction of diatom species despite their tremendous diversity. With the advancement of technology, pelagic diatoms can now be detected via satellite according to remote sensing reflectance based on the diagnostic wavelengths of diatoms (Bracher et al., 2009; Xi et al. 2020). This has allowed quantification of blooms in both spatial and temporal scales that has been unachievable via conventional methods. The application of satellite detection of diatoms provides crucial information of phytoplankton biomass, which are the food source to organisms in the higher trophic levels such as fish. This information will in turn be utilized to assist in the management of fisheries and the projection of future change.

Coastal regions and estuaries are incredibly productive in nutrient recycling and storage of organic matter. These ecosystems provide vital ecosystem services such as carbon fixation, coastal protection, fisheries, habitat diversity, and nutrient removal. Benthic diatom photosynthetic activity is the primary fixed carbon source in these shallow ecosystems, mainly tidal flats (Karsten et al., 2021). Apart from anthropogenic influences and environmental changes such as human activities and pollutants, benthic diatoms are also likely to be affected by climate change and sea temperatures (Pniewski and Piasecka-Jędrzejak, 2020). Furthermore, it is predicted that the impact of climate change on diatoms is likely to intensify soon, with the intensity varying from polar to tropical regions (Malviya et al., 2016; Raven and Beardall, 2021; Rost et al., 2008).

Benthic diatoms are divided into two groups according to their colonization substratum; epipelon and epipsammon (Round et al., 1990). The epipelon consists of free-living diatoms that can move through muddy sediments. In contrast, the epipsammon comprises smaller, non-motile diatoms attached to the particles of sandy sediments. Benthic diatoms are often well preserved in sediments and have been a reliable indicator of environmental variations (Andrén and Jarlman, 2008; Desrosiers et al., 2013). The benthic diatom is well adapted to the highly dynamic intertidal environment, where nutrients, temperature, and light change in response to the tidal change. Sediment type, temperature, light, and nutrients often influence the diatom communities (Costa-Böddeker et al., 2017). Hence, benthic diatoms have carried out different mechanisms in response to the dynamic intertidal environment to ensure survival (Jesus et al., 2009).

6.2.2 Diatoms as Primary Producers

As one of the dominant microalgae and primary producers in the aquatic environment, diatoms exhale about 20–40% of oxygen during photosynthesis. This amount is larger than the oxygen produced by all the rainforests in the world. The aggregate biomass of phytoplankton amounts to less than 1% of the total plant biomass on Earth. Despite the low biomass, phytoplankton are responsible for nearly half of the net primary production. About 90% of the carbon fixation in the ocean is the result of photosynthetic activity executed by phytoplankton. Annually, phytoplankton fixes about 45 gigatons of organic carbon, which is equivalent to the global terrestrial primary production (Behrenfeld et al., 2001; Falkowski et al., 1998). Plants in the marine ecosystems have a rapid turnover time of about 0.02–0.06 years compared to 9–20 years in terrestrial ecosystems. Phytoplankton also act as the biological pump that removes carbon from the ocean surface by sinking fixed organic matter, playing a significant role in regulating the ocean climate (Hinder et al., 2012). Oceanic phytoplankton productivity fluctuations are linked to changes in the network of biogeochemical cycles. This in turn strongly influences the atmospheric CO_2 level on geological time scales. Consequently, we can predict the oceanic biota response in the coming decades by how it is currently influenced by recent ocean phytoplankton productivity, natural climatic variability, and anthropogenic-influenced changes.

Diatoms constitute one of the most diverse and ecologically important phytoplankton groups. They are one of the major primary producers in the sunlit layer of the ocean with abundant nutrients such as the coastal waters and nutrient-rich oceanic regions (e.g., upwelling areas) (Malviya et al., 2016). As the most abundant group in the ocean and benthic ecosystem and because of their high efficiency in soaking up available nutrients, diatoms contribute almost 40% of the total oceanic primary production (Amin et al., 2012). This amount translated to about 20% of global primary productivity (Falkowski et al., 1998), a tremendous amount despite being a microscopic organism that is usually invisible to the naked eye.

In the nutrient-rich (nitrogen (N), phosphorus (P), iron (Fe), and silicon (Si)) region, these ubiquitous algae could multiply and increase productivity with sufficient light as the growth rate and photosynthetic capacity of diatoms are less affected by light fluctuations. In contrast, nutrient limitations impair cellular processes like photosynthesis in most diatoms (Yuan et al., 2019). However, with the oceans constantly changing due to increased atmospheric CO_2 and pH changes (lowering) in diatom-dominated productivity, their role in future phytoplankton communities will be altered (Armbrust, 2009). However, diatoms have proven to have the ability to survive a wide variety of conditions, such as high light variability (Domingues et al., 2011) and elevated CO_2 (Gao and Campbell, 2014). As a result, diatoms are the most successful eukaryotic group by have the ability to acclimate to various environmental changes. Nevertheless, it should be noted that responses to environmental changes are species specific; some diatom species might benefit, and others might not (Wu et al., 2015).

Similarly, in the benthic ecosystem, diatoms constitute an important part of the microphytobenthos community (Karsten et al., 2021). They are considered the most important primary producers responsible for up to 30% of carbon fixation in many intertidal flats (Cahoon and Cooke, 1992; Migné et al., 2009) and are the primary food source for many organisms (Herman et al., 2000). Benthic productivity can surpass the phytoplankton productivity in coastal ecosystems and continental shelf regions. The global primary production of marine phytoplankton is generally

1.8×1016 g C yr^{-1}. In comparison, global production by benthic macrophytes is estimated to be only 5.4×1014 g C yr^{-1} or 3% of the total production of marine phytoplankton (Ackleson, 2003). Likewise, both nutrient and light levels are strongly associated with benthic primary productivity. Unlike their planktonic species, most benthic diatoms are motile and can migrate through the sediment, enhancing light and nutrients (Consalvey et al., 2004; Jauffrais et al., 2016; Vieira et al., 2013) for photosynthesis. By migrating into the sediment, the diatom can avoid photoinhibition due to excessive light (Salleh and McMinn, 2021), which may decrease the primary productivity by 5–30% (Savelli et al., 2021). However, enhanced temperature beyond optimal can reduce the diatom mobility and thus their capability to avoid photoinhibition (Lavaud et al., 2007). Vertical migration provides behavioral photoprotection by moving up and down the sediment to ensure optimal light for photosynthesis and avoid photoinhibitory light levels. More significantly, various factors can influence benthic productivity simultaneously, resulting in a complex in situ variations in benthic primary production.

6.2.3 Diatoms and Nitrogen Fixing

The oceanic nitrogen (N_2) fixation process is essential in providing fixed nitrogen (N) to primary producers and surface water communities (Pajares and Ramos, 2019). Such a pivotal process, which can be further divided into N_2 fixation, nitrification, assimilation, anammox, and denitrification, is mediated mainly by various microorganisms (Voss et al., 2013). However, recent research has also highlighted the involvement of a unique group of open ocean diazotrophs in oceanic N_2 fixation, that is, the heterocystous cyanobacteria that live in symbiosis with primarily diatoms (Foster et al., 2011). Specifically, filamentous N_2 fixing cyanobacteria, such as *Richelia intracellularis* and *Calothrix rhizosoleniae,* are known to form unique and highly specific relationships with several genera of eukaryotic diatoms, including *Hemiaulus*, *Rhizosolenia*, and *Chaetoceros* (Foster et al., 2022; Villareal, 1991).

In addition to providing the required N for N-fixing bacteria, the complex symbiosis relationship between cyanobacterial-diatom benefits diatom counterparts by supplying them directly with the fixed N because in general, diatoms are unable to obtain N from N_2 and thus are dependent on the extracellular dissolved fixed inorganic N pools that are present in very low concentrations in the open ocean (Foster et al., 2011). The symbiotic relationship allows the promotion of growth in both diatoms and the cyanobacterial partners. For example, Foster et al. (2011) demonstrated that free-living cyanobacteria *R. intracellularis* and *C. rhizosoleniae* cells exhibited reduced growth rates compared to those living in symbiosis with the diatom partners, with a significantly 170–420 times higher N_2 fixation rate in cyanobacterial-diatom symbionts compared to free-living *R. intracellularis* and *C. rhizosoleniae* cells.

Cyanobacterial-diatom associations are widely reported in oligotrophic waters and are capable of forming episodic and monospecific blooms with high carbon and nitrogen fixation rates across the globe (Foster et al., 2007). Collectively, they are estimated to fix 4.79 Tg N per year, accounting for approximately a quarter of the total nitrogen input to the ocean (Carpenter et al., 1999). Owing to their N fixing ability, cyanobacterial-diatom associations play a critical role in oceanic nitrogen and carbon cycles by sequestering carbon to the deep sea via aggregation and sinking (Karl et al., 2012). Physical parameters such as light and nutrients are important factors that could affect the nitrogen fixation and growth rates of the host diatom and its diazotrophic endosymbiont. Both growth rates and nitrogen fixation followed light saturation kinetics (Pyle et al., 2020).

The cyanobacterial-diatom symbiotic relationship is unique, whereby each symbiotic filament of diatom is made up of vegetative cells and a terminal heterocyst for N_2 fixation (Caputo et al., 2018). Additionally, a less documented diatom diazotroph association was observed between the pennate diatom, *Climacodium frauenfeldianum*, and cyanobacterial symbiont found in the tropical Atlantic and Pacific Oceans (Carpenter and Janson, 2003). Recently, however, Schvarcz et al. (2022) reported the widespread presence of pennate diatom-cyanobacteria symbioses in the open ocean, namely *Rhopalodiaceae* diatoms *Epithemia pelagica* sp. nov. and *Epithemia catenata* sp. nov. found in the subtropical North Pacific Ocean. Both *E. pelagica* and *E. catenata* symbioses exhibited a unique daily N_2 fixation pattern; unlike other marine cyanobacteria whereby N_2 fixation occurs either during the day or the night, the N_2 fixation of *E. pelagica* and *E. catenata* symbioses occurs during the day and night, with the fixation process stopping either a few hours before the end of the light period in *E. pelagica* or in conjunction with the end of the light period in *E. catenata* (Schvarcz et al., 2022). However, as the increase in anthropogenic CO_2 lowers the oceanic pH and results in ocean acidification, it also inevitably negatively affected the nitrogen fixation rate and growth of cyanobacteria (Shi et al., 2012); it is highly possible that the increase in pH due to ocean acidification might also affect the nitrogen fixing process of cyanobacteria-diatom symbioses.

6.3 BENEFICIAL AQUATIC ROLES OF DIATOMS

Diatoms have been used in many applications such as biofuel exploration, forensic examination, environmental indication, and toxicity testing, and they are also used in the ecosystem eutrophication process (Jamali et al., 2012). Silica and lipid from diatoms are the two major raw materials for industrial applications (Sharma et al., 2021). Diatom-based silica is used for a wide range of biomedical purposes such as analytical tools, drug-delivery systems, nano-particles, biosensors, and nano-materials. Diatom-based lipid is a high-quality feedstock for biofuel production. As a primary producer, diatoms are capable in de novo biosynthesis of the long chain poly-unsaturated fatty acid (LC-PUFA) (Tanaka et al., 2022). Therefore, many diatom species are rich in anti-oxidative and antibacterial compounds, which can be exploited for nutraceutical purposes. Three LC-PUFAs – eicosapentaenoic acid (EPA), docosahexaenoic acid (DHA) and arachidonic acid (ARA) – are precursors for eicosanoids, which are integral for an array of vital cellular activities, including membrane integrity, immune response, gene regulation, and renal and neural functions (Monroig et al., 2022). This makes diatoms an excellent natural source of essential fatty acids for humans and livestock. Peter et al. (2021) reported that diatoms can be used as an alternative for the biological treatment of wastewater because of their ability to take up or consume pollutants for their growth. Thus, in the following sections we will be focusing on the roles of diatoms in the aquatic environment, their applications, and the part they play in benefitting humans.

6.3.1 Diatoms as Environmental and Biochemical Markers

With the rapid change in environmental factors due to climate change, it is essential for us to document the consequential impact on the population assemblage of diatoms, so that the indices used will accurately reflect the desired monitoring capacity. Benthic diatoms have been employed as important aquatic environmental assessment indicators for the past century (Kolkwitz and Marsson, 1908) and are now part of the standard bioassessment toolkits used worldwide (Charles et al., 2021). For example, diatoms are part of the assessment criteria for water bodies in accordance with the United States Clean Water Act (1972) or the European Water Framework Directive. Although there are now various diatom indices, including acidification and salinization, the most commonly used index is still the eutrophication indicator, especially in the freshwater environment where eutrophication is being considered as the most common stressor (Bellinger et al., 2006; Poikane et al., 2021). Diatoms are capable of being used as reliable indicators of the water quality owing to their sensitivity toward changes in their aquatic environment.

Additionally, due to their fast reproduction rate, a given species can proliferate on an exponential scale under favorable conditions while others concurrently decrease or are displaced, thereby allowing us to determine the changes in any physiochemical conditions. Some of the diatom indices that are being implemented across the globe include IPS (Specific Pollution Sensitivity Index, CEMAGREF, 1982), Trophic Diatom Index (TDI, Kelly and Whitton, 1995), and the Diatom Biological Index (DBI, Coste et al., 2009). In general, the composition or assemblages of diatoms, especially at the genus level, is used for accurate prediction of environmental changes in most assessment studies of the aquatic environment for anthropogenic eutrophication activities (Lu et al., 2020).

Aside from assessing eutrophication activities, diatoms are also able to be used as valuable indicators for heavy metals as they showed great sensitivity toward heavy metal exposure (Masmoudi et al., 2013; Medley and Clements, 1998). For example, some streams and basins are acidic with high levels of dissolved sulphates, toxic metals, and metalloids, subsequently causing acid rock drainage. In addition, mining activities will contribute to acid mine drainage, and together with acid rock drainage, these cause stress toward diatom communities. Therefore, diatom floras, diatom cell size, and growth are generally lower in accordance with the severity of the mining impact and/or heavy metal pollution (Morin et al., 2014). As such, a new diatom-based index that incorporates the Metal Pollution Index was developed by Fernández et al. (2018) to evaluate heavy metal pollution in water bodies and has produced satisfactory results in the Tinto River, Odiel River, and Guadiamar River in Spain.

In addition to acting as general environmental and biochemical markers, owing to their role in organic carbon transport from the ocean surface to the deep sea and their efficiency as a biological pump for carbon dioxide, diatoms can also function as promising biological indicators to quantify past climatic changes such as changes in temperature and other connecting limnological factors, thereby allowing us to predict future climate change conditions (Bigler and Hall, 2002). This is possible because diatoms are among the various proxy-data that is being preserved in good conditions in lake sediments, especially in undisturbed environments or regions such as the polar regions where their composition and productivity are mainly only regulated by climatic conditions (Shukla et al., 2009).

6.3.2 Diatoms for Water Treatment

The unique porous, nanostructured silica cell wall of diatoms can be manipulated and is widely used in the process of waste degradation (Marchetti and Cassar, 2009). Diatoms respond quickly to the presence of organic matter and nutrient contamination like metolachlor, atrazine, simazine, phenols, phosphate, nitrate, urea, iron, and heavy metals such as copper, cadmium, chromium, magnesium, zinc, and lead for their growth, as well as accommodation in the reduction of biochemical oxygen demand (Arora et al., 2021; Blanco and Becares, 2010; Marella et al., 2020). The abundance of diatoms in the water column helps enhance the removal of the pollution from the wastewater originating from the industrial process, from agricultural activities, and from other anthropogenic activities (Adesra et al., 2021; Marella et al., 2020). The algal-bacteria consortium particularly is useful in wastewater treatment as this symbiotic process can produce higher algae biomass and enhance contamination removal, thus helping to reduce the bioremediation cost for polluted wastewater (Khoo et al., 2021). The algal-bacterial consortium removes the pollution from the wastewater system through several process such as assimilation (uptake of nitrogen and phosphorus), stripping (ammonia removal at high pH), nitrification-denitrification, adsorption (heavy metal removal), and phosphorus precipitation and pathogen removal from pH fluctuation (Oruganti et al, 2022). Renuka et al. (2015) recognized the use of microalgae in wastewater remediation as an eco-friendly process, as long as no secondary pollution happens and the biomass produced is being used to allow efficient nutrient recycling.

The application of microalgae in wastewater treatment promotes an environmentally friendly alternative from the conventional method of wastewater treatment using inorganic chemicals. They are efficient in reducing operational costs, and the biomass obtained from the process could be used as feedstock or biomaterial for biorefinery production of lipids and biofuels (Peter et al., 2021; Show et al. 2021; Yong et al., 2021). Microalgae has been used in phycoremediation of domestic wastewater in which the organic waste is rich in nutrients required for microalgae growth. Thus, this results in the excellent removal of nitrogen and phosphorus from the water (Show et al., 2017).

The type of diatom species and its applications in wastewater treatment are shown in Table 6.1. Species from the genus *Nitzschia* are frequently used in wastewater treatment as it possesses high tolerance toward pollutants and assists its biodegradation process (Abdel-Raouf et al., 2012; Hong et al., 2008; Malinowska-Gniewosz et al., 2018; Palmer, 1969; Xing et al. 2007, 2008, 2018). Xing et al. (2008) showed that the diatom *Nitzschia* sp. can be used to treat aquaculture affluents and can be used as feed for abalone, sea urchin, and sea cucumber. Additionally, common diatom species used for aquaculture feed such as *Skeletonema costatum, Chaetoceros calcitrans,* and *Thalassiosira* sp., also have the potential for efficient wastewater treatment, thus paving the way for sustainable aquaculture practices (Hong et al., 2008; Loix et al., 2017; Tiwari and Marella, 2019; Volland et al., 2014).

6.3.3 Diatoms in Aquaculture

Aquaculture is a vital contributor toward sustainable global food security in providing a healthy and cheaper source of protein to feed the growing world population. The exponential growth of aquaculture increases the demand for feed. Adding to that, the unstable volume supply of traditional feed ingredients such as fish oil and soybean meal, due to climate, political, and economic factors, also lead to increasing feed production costs, consequently raising the price of aquaculture products (Dawood et al., 2022). The Food and Agriculture Organization of the United Nations, FAO (2020), has reported that fish oil for aquaculture feed production in 2018 was 18 million tonnes, higher than 14 million tonnes recorded in 2014. This has forced the industry to search for alternative ingredients to partially or fully replace these traditional ingredients in order to meet the

TABLE 6.1
Types of Diatom Species with Their Applications in Wastewater Treatment

Species	Application	Reference
Nitzschia sp., *Navicula* sp.	Tolerance to organic pollutants	Abdel-Raouf et al. (2012), Palmer (1969)
Nitzschia sp.	Aquaculture wastewater treatment and feed for culture species	Xing et al. (2007), Xing et al. (2008)
Skeletonema costatum, Nitzschia sp.	Biodegradation of polycyclic aromatic hydrocarbons (PAHs)-phenanthrone (PHE) and fluoranthene (PLA)	Hong et al. (2008)
Synedra sp., *Synedra tenera, Synedra acus*	Removing turbidity, water treatment	Kim et al. (2014).
Chaetoceros calcitrans	Absorbing heavy metals like Hg	Volland et al. (2014)
Phaeodactylum trcornutum, Thalassiosira weissflogii, Thalassiosira oceanica	Absorbing heavy metal, especially cadmium, Cd metal ion	Loix et al. (2017)
Cymatopleura radiosa, Navicula upsaliensis, Nitzschia angustata	Tolerant to the organic pollution and eutrophication process	Malinowska-Gniewosz et al. (2018)
Nitzschia sp.	Reducing environmental harm of total nitrogen (TN), ammonium nitrogen (NH_4^-N), and total phosphate (TP) in aquaculture wastewater	Xing et al. (2018)
Cyclotella cryptica, Skeletonema costatum, Cyylindrotheca fusiformis, Phaeodactylum tricornutum, Thalassiosira pseudonana	Controlling heavy metal and heavy metal resistance	Tiwari and Marella (2019)

demand of protein and lipid sources for aquaculture feed production worldwide (Bell, 2008; Froehlich et al., 2018; Trushenski and Lochmann, 2009).

6.3.3.1 Feedstock

Agricultural products and by-products (barley, corn, canola, oat, pea, soybean, sunflower, and wheat) are traditionally considered as excellent choices for nutritional components in feed formulation to fulfill the dietary requirements in the livestock industry, both terrestrial and aquatic. Such raw materials are available in abundance and are more cost-reducing than animal-sourced ingredients, such as blood meal, offal meal, bone meal, fishmeal, fish liver oil, and inedible fats (lard and poultry fat) from rendering facilities (Arnesen and Krogdahl, 1993; Gatlin et al., 2007; Kerr et al., 2015). Since terrestrial plants are not a natural food source in the aquatic environment, this has led to low palatability and poor digestibility, raising concerns on environmental nitrogen pollution from feed waste and excretion (Chai et al., 2021).

By right, the best sources of protein and oil for the diets of aquatic organisms are the primary producer itself, which is the diatoms. Diatoms have a balanced nutritional profile, providing a natural source of proteins, lipids, minerals, carbohydrates, and bioactive compounds, such as alginate, carotenoids, and polyphenols (Koyande et al., 2019; López-Pedrouso et al., 2020; Pocha et al., 2022), making them a good replacement for fishmeal and fish oil. However, low feasibility hinders this as the production cost is still too high (Beal et al., 2018). Nonetheless, inclusion of diatoms in aquaculture feed has proven their suitability as an aquafeed ingredient, as shown in Table 6.2. Dietary inclusion of diatom meal improved feed efficiency subsequently shown in weight gain in Nile tilapia, *Oreochromis niloticus* (Ayoub et al., 2019), Atlantic salmon *Salmo salar* L (Sørensen et al., 2016), Olive flounder *Paralichthys olivaceus* (Noh et al., 2020), Channel catfish *Ictalurus punctatus* (Li et al., 2009), and African catfish *Clarias gariepinus* (Raji et al., 2018). The diatoms *Amphora coffeaeformis, Melosira nummuloides,* and *Chlorella vulgaris* boost immunity when supplemented in feed (Ayoub et al., 2019; Noh et al., 2020; Raji et al., 2018; Saleh et al., 2020). Inclusion of Arctic diatom *Porosira glacialis* in the Atlantic salmon *Salmo salar* L. diet interestingly reduced salmon-lice infestation, probably due to the oxidative activity from its free fatty acid content (Eilertsen et al., 2021).

6.3.3.2 Hatchery Technology

A significant hurdle to overcome by the aquaculture industry is the inability to provide enough live food during the nursery phase. The most critical stage for larval production is the late to post-larval stage in important aquaculture species of fish (Bricknell and Dalmo, 2005; Chen et al., 2020; Sorgeloos et al., 2001), penaeid shrimps (Arulmoorthy et al., 2020; Manan and Ikhwanuddin, 2021), mudcrabs (Syafaat et al., 2021), molluscs (Mau and Jha, 2018; Uriarte et al., 2019), and bivalves (Barros et al., 2018; Cheng et al., 2020). Artificial or inert feed has been introduced to encourage early weaning off live food (de Moraes et al., 2022). Nonetheless, it is not as nutrient-efficient as natural live food, such as *Artemia* naupliis, rotifers, copepods, diatoms, and microalgae, in fulfilling larval dietary requirements. Thus, hatchery operators have to still rely on mass cultivation of live food for this purpose. Pure cultures of diatoms, for example, *Chaetoceros, Dunaliella, Isochrysis, Skeletonema, Tetraselmis and Nannochloropis*, are frequently used in larval maintenance (Kaparapu, 2018). A feeding strategy using a single diatom species is less cumbersome for hatchery operations but better production can be achieved when cultured species are fed a mixture of diatoms. Ragg et al. (2010) reported that green mussel *Perna canaliculus* larvae show high survival and good growth when fed 66% C. *calcitrans* + 33% *I. galbana*. A mix of *Navicula* cf. *lenzii* Hustedt and *Amphora luciae* Cholnoky high in omega-3 LC-PUFAs helps to improve survival of the abalone *Haliotis discus hannai* postlarvae, as reported by Gordon et al. (2006). Likewise, 50% *Skeletonema costatum* + 50% *Rhodomonas marina* feeding improves the survival rate of purple sea urchin *Paracentrotus lividus* for up to 50% in the early larvae stages (Araújo et al., 2020).

TABLE 6.2
Inclusion of Diatoms in the Feed of Important Aquaculture Species

Cultured Species	Diatom Species (Meal form)	Inclusion in Feed (as Fed Basis)	Effects	Reference
Nile tilapia (*Oreochromis niloticus*)	*Amphora coffeaeformis*	1% to 2%	Increases weight gain, improves protein and lipid muscle content, boosts non-specific immunity	Ayoub et al. (2019)
	Amphora coffeaeformis	7.5%	Boosts immunity, enhances male maturity	Saleh et al. (2020)
	Nanofrustulum sp.	40.12%	Similar growth with fishmeal and displays no negative effects, suitable as an alternative ingredient	García-Ortega et al. (2015)
Atlantic salmon (*Salmo salar* L.)	*Porosira glacialis*	2%	Significant lice deterrent	Eilertsen et al. (2021)
	Entomoneis spp.	2.5% to 5%	Similar growth with control feed and displays no negative effects, increases whole-body n-3 content	Norambuena et al. (2015)
	Phaeodactylum tricornutum	3% to 6%	Increases weight gain, improves whole-body LC-PUFA content	Sørensen et al. (2016)
Olive flounder (*Paralichthys olivaceus*)	*Melosira nummuloides*	1% or 2% *M. nummuloides* + 1% *Lactobacillus plantarum*	Increases weight gain, boosts non-specific immunity, displays low expression of immune-related genes (galectin and lysosome)	Noh et al. (2020)
Channel catfish (*Ictalurus punctatus*)	*Schizochytrium* sp.	2%	Increases weight gain, improves whole-body n-3 content	Li et al. (2009)
African catfish (*Clarias gariepinus*)	*Chlorella vulgaris*	12.5% to 18.75%	Improves growth performance, improves red and white cell counts, decreases total cholesterol and plasma glucose levels	Raji et al. (2018)

Aside from their nutritional benefits, diatoms are also used in improving nursery techniques for aquatic invertebrate culture. In commercial abalone hatcheries, for example, the application of biofilms consisting of a mixture of bacteria and diatoms embedded in a mucus-like matrix is used to induce abalone larvae settlement instead of relying on non-geniculate coralline red algae (NCA) in their natural environment (Daume et al., 2004; Daume, 2006). The biofilm consisting of macroalgae *Ulvella lens* and diatom *Amphora* sp. has been successfully applied in inducing the abalone larvae settlement of various abalone species such as *H. discus hannai* and *Haliotis rubra* (Hannon et al., 2014), whereas *Haliotis laevigata* prefers settlement on a biofilm consisting of NCA *Sporolithon durum* with the diatom *Navicula ramosissima* (Daume et al. 1999).

6.3.3.3 Health and Disease Management

One of the health management strategies in aquaculture is biocontrol of pathogenic aquaculture bacterias, particularly *Vibrio* sp. by co-culturing diatoms with targeted aquaculture species. Laboratory studies showed that several species of diatoms significantly regulated the growth of vibrios. The inhibitory effect of six strains of benthic diatoms, *Nitzschia laevis* Hustedt, *Nitzschia frustulum* var. *perminuta* Grunow, *Navicula incerta* Grunow, *Navicula biskanterae* Hust., *Nitzschia fustulum* (Kútzing) Grunow, and *Navicula* cf. *incerta* Grunow, were assessed against three species of vibrios, *Vibrio alginolyticus*, *V. campbellii*, and *V. harveyi* (Molina-Cárdenas and del Pilar Sánchez-Saavedra, 2017). All diatoms significantly reduced all bacterial loads for all vibrio species. In addition, purified antibacterial compounds of *Skeletonema costatum* (Grev.) Cleve extract show an antagonistic effect against *Vibrio anguillarum* with significant inhibition of its growth (Naviner et al., 1999). This strategy can be applied in reducing or replacing the use of antibiotics in aquaculture.

In terrestrial livestock, Nazmi et al. (2017) also proposed the use of diatoms as a safer vaccine adjuvant in inactivated vaccines for broilers. Common adjuvants like aluminium hydroxide often produced inflammation at the injection site, which consequently degraded the meat quality. In their study, efficacy of adjuvant prepared from *Thalassiosira pseudonana* (CCMP1335) mixed with inactivated Newcastle Disease Virus were tested on chickens. The results are comparable to aluminium hydroxide and commercially available Freund's incomplete adjuvant, which indicate the potential for diatoms to act as an alternative adjuvant. This method can potentially be administrated in vaccinations of aquatic mammals in captivity like dolphins and orcas, and may also be applied to larger fish species like catfish and grouper.

6.4 FUTURE FOR DIATOMS AND HOW THEIR CURRENT ROLE MIGHT CHANGE

In the foreseen future, diatoms could well be the primary frontrunner for quality raw materials for sustainable food security, as feedstock, and also for direct human consumption. As such, a feasibility study is needed to evaluate and promote the use of diatom by-products from biorefineries so that they are cost-competitive to be a value-added feed ingredient. Active involvement between global regulatory bodies (e.g., International Energy Agency (IEA)) with the FAO is needed to improve international cooperation and information exchange between bioenergy research and food security programs in different countries.

Climate change is predicted to have dramatic effects on marine ecosystems. Since the Industrial Revolution, burning of fossil fuels and other anthropogenic activities have released ~100 times more CO_2 above the natural levels into the atmosphere (Berner, 2003). It is estimated that about 30% of atmospheric CO_2 is absorbed by the oceans (Sabine et al., 2004), resulting in more available dissolved CO_2 in the oceans. This has caused the unwanted side effect of CO_2 emission in the ocean called ocean acidification (Caldeira and Wickett, 2003). Hence, it is critical to understand what the future lies ahead for diatoms and how diatoms will respond to changes as any changes to diatoms will have a domino effect on the aquatic food web and the fluxes of elements in the oceans (Armburst et al., 2009). Rising CO_2 can influence the growth and productivity of diatoms. Previous studies have projected that phytoplankton might grow faster with more CO_2. However, a recent study shows that at certain nutrient levels, diatoms stopped harvesting as much energy from sunlight and carried out less respiration (Hennon et al., 2015). The studied diatom species was *Thalassiosira pseudonana*, which has shown reduced transcription of photosynthesis and respiration in response to sudden or long-term increases in CO_2. Diatoms have also shown resiliency under low pH conditions where they can sustain their growth under high CO_2 conditions than in a low CO_2 environment (Valenzuela et al., 2018). Ocean warming is opening up areas that were previously occupied by sea ice and prolonged the open water season in the Arctic Ocean. This has resulted in more light penetrating the surface ocean, and satellite data shows that the total annual primary productivity in the Arctic Ocean has increased about 10–20% over the period from 1998–2010 (Belanger et al, 2013). With the projection of more opening of ice-free areas in the Arctic and Antarctic, this could further increase diatom productivity and may in some way mitigate CO_2 emission to a certain degree. This shows that it is not all doom and gloom for diatoms, and there could be a multitude of pathways for diatoms in the future.

6.5 CONCLUSION

Diatoms play an essential role, both in ensuring the normal energy and nutrient cycles, and in serving as a primary food source for other organisms in the food chain. Owing to their important role and their sensitivity toward almost all environmental factors, including land use change and water quality, diatom indices are widely being applied globally to aid in the monitoring, assessment, and management of both freshwater and marine water environments. Active involvement of governmental organizations or authorized global regulatory bodies are needed to standardize production at the national level or worldwide. This is in the hope that they can overcome issues in standardizing/optimizing the industrial protocols for efficient production toward the development of diatom mass production and thus strengthen the concept that diatom resources are sustainable options.

REFERENCES

Abdel-Raouf, N., A. A. Al-Homaidan and I. B. M. Ibraheem. 2012. Microalgae and Wastewater treatment. *Saudi Journal of Biological Science*, 19(3): 257–275. 10.1016/j.sjbs.2012.04.005

Ackleson, S. G. 2003. Light in shallow waters: A brief research review. *Limnology and Oceanography*, 48(1 part II): 323–328. 10.4319/lo.2003.48.1_part_2.0323

Adesra, A., V. K. Srivastava and S. Varjani. 2021. Valorization of Dairy Wastes: Integrative Approaches for Value Added Products. *Indian Journal of Microbiology*, 61(3): 270–278.

Agustí, S., J. I. González-Gordillo, D. Vaqué, M. Estrada, M. I. Cerezo, G. Salazar, J. M. Gasol, and C. M. Duarte. 2015. Ubiquitous healthy diatoms in the deep sea confirm deep carbon injection by the biological pump. *Nature Communications*, 6(1): 1–8. 10.1038/ncomms8608

Amin, S. A., M. S. Parker and E. V. Armbrust. 2012. Interactions between Diatoms and Bacteria. *Microbiology and Molecular Biology Reviews*, 76(3): 667–684. 10.1128/mmbr.00007-12

Andrén, C., and A. Jarlman. 2008. Benthic diatoms as indicators of acidity in streams. *Fundamental and Applied Limnology*, 173(3): 237–253. 10.1127/1863-9135/2008/0173-0237

Araújo, J., A. Candeias-Mendes, I. Monteiro, D. Teixeira, F. Soares, and P. Pousão-Ferreira. 2020. The use of diatom *Skeletonema costatum* on aquaculture-produced purple sea urchin (*Paracentrotus lividus*) larvae and post-larvae diet. *Aquaculture Research*, 51(6): 2545–2554. 10.1111/are.14597

Armbrust, E. V. 2009. The life of diatoms in the world's oceans. *Nature*, 459(7244): 185–192. 10.1038/nature08057

Arora, K., P. Kaur, P. Kumar, A. Singh, S. K. S. Patel, X. Li, Y. H. Yang, S. K. Bhatia, and S. Kulshrestha. 2021. Valorization of Wastewater resources into biofuel and Value-Added Products using microalgal System. *Frontiers in Energy Research*, 9: 1–25. 10.3389/fenrg.2021.646571

Arnesen, P., and Krogdahl, Å. (1993). Crude and pre-extruded products of wheat as nutrient sources in extruded diets for Atlantic salmon (Salmo salar, L) grown in sea water. *Aquaculture*, 118, 105–117. 10.1016/0044-8486(93)90284-6.

Arulmoorthy, M. P., E. Anandajothi, S. Vasudevan, and E. Suresh. 2020. Major viral diseases in culturable penaeid shrimps: a review. *Aquaculture International*, 28(5): 1939–1967. 10.1007/s10499-020-00568-3

Ayoub, H. F., M. F. Abdelghany and A. B. El-Sayed. 2019. Effects of Diatoms *Amphora coffeaeformis* on growth parameters, nonspecific immunity and protection of the Nile tilapia (*Oreochromis niloticus*) to *Aeromonas hydrophila* infection. *Egyptian Journal of Aquatic Biology and Fisheries*, 23: 413–426. 10.21608/ejabf.2019.37466

Barros, J., F. M. Winkler and L. A. Velasco. 2018. Heritability, genetic correlations and genotype-environment interactions for growth and survival of larvae and post-larvae of the Caribbean scallop, *Argopecten nucleus* (Mollusca: Bivalvia). *Aquaculture*, 495: 948–954. 10.1016/j.aquaculture.2018.06.047

Beal, C. M., L. N. Gerber, S. Thongrod, W. Phromkunthong, V. Kiron, J. Granados, I. Archibald, C. H. Greene and M. E. Huntley. 2018. Marine microalgae commercial production improves sustainability of global fisheries and aquaculture. *Scientific Reports*, 8(1): 1–8. 10.1038/s41598-018-33504-w

Belanger, S., M. Babin and J.-E. Tremblay. 2013. Increasing cloudiness in arc- tic damps the increase in phytoplankton primary production due to sea ice receding. *Biogeosciences*, 10(6): 4087–4101. 10.5194/bg-10-4087-2013

Bellinger, B. J., C. Cocquyt and C. M. O'Reilly. 2006. Benthic diatoms as indicators of eutrophication in tropical streams. *Hydrobiologia*, 573: 75–87. 10.1007/s10750-006-0262-5

Bell, J. G. 2008. Alternative lipid sources for fish feed and farmed fish quality. In *Improving farmed fish quality and safety*, ed.Ø. Lie, 275–299. Woodhead Publishing. 10.1533/9781845694920.2.275

Berner, R. A. 2003. The long-term carbon cycle, fossil fuels and atmospheric composition. *Nature*, 426: 323–326. 10.1038/nature02131

Bigler, C. and R. I. Hall. 2002. Diatoms as indicators of climatic and limnological change in Swedish Lapland: a 100-lake calibration set and its validation for paleoecological reconstructions. *Journal of Paleolimnology*, 27: 97–115. 10.1023/A:1013562325326

Blanco, S. and E. Becares. 2010. Are biotic indices sensitive to river toxicants? A comparison of metrics based on diatoms and macro-invertebrates. *Chemosphere*, 79: 18–25. 10.1016/j.chemosphere.2010.01.059

Bracher, A., M. Vountas, T. Dinter, J. P. Burrows, R. Roettgers, and I. Peeken. 2009. Quantitative observation of cyanobacteria and diatoms from space using PhytoDOAS on SCIAMACHY data. *Biogeosciences*, 6(5):751–764. 10.5194/bg-6-751-2009

Behrenfeld, M., J. Randerson, C. McClain, G. Feldman, S. Los, C. Tucker, P. Falkowski, C. Field, R. Frouin, W. Esaias and D. D. Kolber. 2001. Biospheric primary production during an ENSO transition. *Science*, 291(5513): 2594–2597. 10.1126/science.1055071

Bricknell, I. and R. A. Dalmo. 2005. The use of immunostimulants in fish larval aquaculture. *Fish and Shellfish Immunology*, 19(5): 457–472. 10.1016/j.fsi.2005.03.008

Cahoon, L. and J. Cooke. 1992. Benthic microalgal production in Onslow Bay, North Carolina, USA. *Marine Ecology Progress Series*, 84(2): 185–196. https://www.jstor.org/stable/24829553

Caldeira, K. and M. Wickett. 2003. Anthropogenic carbon and ocean pH. *Nature*, 425: 365. 10.1038/425365a

Caputo, A., M. Stenegren, M. C. Pernice and R. A. Foster. 2018. A short comparison of two marine planktonic diazotrophic symbioses highlights an un-quantified disparity. *Frontiers in Marine Science*, 5: 2. 10.3389/fmars.2018.00002

Carpenter, E. J. and S. Janson. 2003. Intracellular cyanobacterial symbionts in the marine diatom *Climacodium frauenfeldianum* (Bacillariophyceae). *Journal of Phycology*, 36: 540–544. 10.1046/j.1529-8817.2000.99163.x

Carpenter, E. J., J. P. Montoya, J. Burns, M. R. Mulholland, A. Subramaniam, and D. G. Capone. 1999. Extensive bloom of a N_2-fixing diatom/cyanobacterial association in the tropical Atlantic Ocean. *Marine Ecology Progress Series*, 185: 273–283. 10.3354/meps185273

Cedhagen, T., W. Cheah, A. Bracher and F. Lejzerowicz. 2014. Algal pigments in Southern Ocean abyssal foraminiferans indicate pelagobenthic coupling. *Deep Sea Research Part II: Topical Studies in Oceanography*, 108: 27–32. 10.1016/j.dsr2.2014.07.017

CEMAGREF. 1982. Etude des méthodes biologiques d'appréciation quantitative de la qualité des eaux. In *Bassin Rhône-Mediterrannée-Corse*, eds Q. E. Rapport and A. F. Lyon, 218 Lyon: CSIRO.

Charles, D. F., M. G. Kelly, R. J. Stevenson, S. Poikane, S. Theroux, A. Zgrundo and M. Cantonati. 2021. Benthic algae assessments in the EU and the US: striving for consistency in the face of great ecological diversity. *Ecological Indicators*, 121: 107082. 10.1016/j.ecolind.2020.107082

Chai, W. S., W. G. Tan, H. S. H. Munawaroh, V. K. Gupta, S. H. Ho and P. L. Show. 2021. Multifaceted roles of microalgae in the application of wastewater biotreatment: a review. *Environmental Pollution*, 269: 116236. 10.1016/j.envpol.2020.116236

Chen, J. Y., C. Zeng, D. R. Jerry, and J. M. Cobcroft. 2020. Recent advances of marine ornamental fish larviculture: broodstock reproduction, live prey and feeding regimes, and comparison between demersal and pelagic spawners. *Reviews in Aquaculture*, 12(3): 1518–1541. 10.1111/raq.12394

Cheng, P., C. Zhou, R. Chu, T. Chang, J. Xu, R. Ruan, P. Chen and X. Yan. 2020. Effect of microalgae diet and culture system on the rearing of bivalve mollusks: Nutritional properties and potential cost improvements. *Algal Research*, 51: 102076. 10.1016/j.algal.2020.102076

Consalvey, M., B. Jesus, R. G. R. Perkins, V. Brotas, G. J. C. Underwood and D. M. Paterson. 2004. Monitoring migration and measuring biomass in benthic biofilms: The effects of dark/far-red adaptation and vertical migration on fluorescence measurements. *Photosynthesis Research*, 81: 91–101. 10.1023/B:PRES.0000028397.86495.b

Costa-Böddeker, S., L. X. Thuyên, A. Schwarz, H. Đ. Huy and A. Schwalb. 2017. Diatom Assemblages in Surface Sediments Along Nutrient and Salinity Gradients of Thi Vai Estuary and Can Gio Mangrove Forest, Southern Vietnam. *Estuaries and Coasts*, 40(2): 479–492. 10.1007/s12237-016-0170-5

Coste, M., S. Boutry, J. Tison-Rosebery and F. Delmas. 2009. Improvements of the Biological Diatom Index (BDI): description and efficiency of the new version (BDI-2006). *Ecological Indicators*, 9: 621–650. 10.1016/j.ecolind.2008.06.003

Daume, S., S. Huchette, S. Ryan and R. W. Day. 2004. Nursery culture of *Haliotis rubra*: the effect of cultured algae and larval density on settlement and juvenile production. *Aquaculture*, 236(1–4): 221–239. 10.1016/j.aquaculture.2003.09.035

Daume, S. 2006. The roles of bacteria and micro and macro algae in abalone aquaculture: a review. *Journal of Shellfish Research*, 25(1): 151–157. 10.2983/0730-8000(2006)25[151:TROBAM]2.0.CO;2

Daume, S., Brand-Gardner, S., and Woelkerling, W.J (1999). Preferential settlement of abalone larvae: diatom films vs. non-geniculate coralline red algae. *Aquaculture*, 174, 243–254 10.1016/s0044-8486(99)00003-4.

Dawood, M. A. O., Habotta, O. A. E., Elsabagh, M., Azra, M. N., Van Doan, H., Kari, Z. A., and Sewilam, H. (2022). Fruit processing by-products in the aquafeed industry: A feasible strategy for aquaculture sustainability. *Reviews in Aquaculture*, 14, 1945–1965 10.1111/raq.12680.

de Moraes, L. B. S., R. F. B. Santos, G. F. Gonçalves Junior, G. C. P. Mota, D. M. D. M. Dantas, R. de Souza Bezerra and A. Olivera Gálvez. 2022. Microalgae for feeding of penaeid shrimp larvae: an overview. *Aquaculture International*, 30: 1295–1313. 10.1007/s10499-022-00857-z

Deppeler, S. L. and A. T. Davidson. 2017. Southern Ocean phytoplankton in a changing climate. *Frontiers in Marine Science*, 4: 40. 10.3389/fmars.2017.00040

Desrosiers, C., J. Leflaive, A. Eulin, and L. Ten-Hage. 2013. Bioindicators in marine waters: Benthic diatoms as a tool to assess water quality from eutrophic to oligotrophic coastal ecosystems. *Ecological Indicators*, 32: 25–34. 10.1016/j.ecolind.2013.02.021

Domingues, R. B., T. P. Anselmo, A. B. Barbosa, U. Sommer and H. M. Galvão. 2011. Light as a driver of phytoplankton growth and production in the freshwater tidal zone of a turbid estuary. *Estuarine, Coastal and Shelf Science*, 91(4): 526–535. 10.1016/j.ecss.2010.12.008

Eilertsen, H. C., E. Elvevoll, I. H. Giæver, J. B. Svenning, L. Dalheim, R. A. Svalheim, B. Vang, S. Siikavuopio, R. Dragøy, R. A. Ingebrigtsen and E. Hansen. 2021. Inclusion of photoautotrophic cultivated diatom biomass in salmon feed can deter lice. *PloS one*, 16(7): e0255370. 10.1016/j.ecss.2010.12.008

Falkowski, P. G., R. T. Barber and V. Smetacek. 1998. Biogeochemical controls and feedbacks on ocean primary production. *Science*, 281(5374): 200–206. 10.1126/science.281.5374.200

FAO. (2020). The State of World Fisheries and Aquaculture 2020. Sustainability in action. In *FAO*. 10.4060/ca9229en

Fernández, M. R., G. Martín, J. Corzo, A. de la Linde, E. García, M. López and M. Sousa. 2018. Design and testing of a new diatom-based index for heavy metal pollution. *Archives of Environmental Contamination and Toxicology*, 74: 170–192. 10.1007/s00244-017-0409-6

Foster, R. A., M. M. M. Kuypers, T. Vagner, R. W. Paerl, N. Musat and J. P. Zehr. 2011. Nitrogen fixation and transfer in open ocean diatom-cyanobacterial symbioses. *The ISME Journal*, 5: 1484–1493. 10.1038/ismej.2011.26

Foster, R. A., A. Subramaniam, C. Mahaffey, E. J. Carpenter, D. G. Capone and J. P. Zehr. 2007. Influence of the Amazon River plume on distributions of free-living and symbiotic cyanobacteria in the western tropical north Atlantic Ocean. *Limnology and Oceanography*, 52: 517–532. 10.4319/lo.2007.52.2.0517

Foster, R. A., D. Tienken, S. Littmann, M. J. Whitehouse, M. M. M. Kuypers, A. E. White. 2022. The rate and fate of N_2 and C fixation by marine diatom-diazotroph symbioses. *The ISME Journal*, 16: 477–487. 10.1038/s41396-021-01086-7

Froehlich, H. E., C. A. Runge, R. R. Gentry, S. D. Gaines and B. S. Halpern. 2018. Comparative terrestrial feed and land use of an aquaculture-dominant world. *Proceedings of the National Academy of Sciences of the United States of America*, 15(20): 5295–5300. 10.1073/pnas.1801692115

García-Ortega, A., L. Martínez-Steele, D. Gonsalves, M. M. Wall and P. J. Sarnoski. 2015. Use of biofuel by-product from the green algae *Desmochloris* sp. and diatom *Nanofrustulum* sp. meal in diets for nile tilapia *Oreochromis niloticus*. *Journal of Aquaculture Engineering and Fisheries Research*, 1(4): 144–154.

Gameiro, C., J. Zwolinski and V. Brotas. 2011. Light control on phytoplankton production in a shallow and turbid estuarine system. *Hydrobiologia*, 669(1): 249–263. 10.1007/s10750-011-0695-3

Gao, K., and D. A. Campbell. 2014. Photophysiological responses of marine diatoms to elevated CO2 and decreased pH: A review. *Functional Plant Biology*, 41(5): 449–459. 10.1071/FP13247

Gatlin, D. M., F. T. Barrows, P. Brown, K. Dabrowski, T. G. Gaylord, R. W. Hardy, E. Herman, G. Hu, Å. Krogdahl, R. Nelson, K. Overturf, M. Rust, W. Sealey, D. Skonberg, E. J. Souza, D. Stone, R. Wilson and E. Wurtele. 2007. Expanding the utilization of sustainable plant products in aquafeeds: a review. *Aquaculture Research*, 38(6): 551–579. 10.1111/j.1365-2109.2007.01704.x

Gordon, N., A. Neori, M. Shpigel, J. Lee and S. Harpaz. 2006. Effect of diatom diets on growth and survival of the abalone *Haliotis discus hannai* postlarvae. *Aquaculture*, 252(2–4): 225–233. 10.1016/j.aquaculture.2005.06.034

Hannon, C., R. A. Officer, J. Le Dorven, and J. Chamberlain. 2014. Culture methods of live algal feeds for European aquaculture: optimising culture conditions for *Ulvella lens*. *Aquaculture International*, 22(6): 1813–1822. 10.1007/s10499-014-9784-4

Hennon, G. M., J. Ashworth, R. D. Groussman, C. Berthiaume, R. L. Morales, N. S. Baliga, M. V. Orellana and E. V. Armbrust. 2015. Diatom acclimation to elevated CO2 via cAMP signalling and coordinated gene expression. *Nature Climate Change* 5(8): 761–765. 10.1038/nclimate2683

Herman, P. M. J., J. J. Middelburg, J. Widdows, C. H. Lucas and C. H. R. Heip. 2000. Stable isotopes as trophic tracers: Combining field sampling and manipulative labelling of food resources for macrobenthos. *Marine Ecology Progress Series*, 204: 79–92. 10.3354/meps204079

Hinder, S. L., G. C. Hays, M. Edwards, E. C. Roberts, A. W. Walne and M. B. Gravenor. 2012. Changes in marine dinoflagellate and diatom abundance under climate change. *Nature Climate Change*, 2(4): 271–275. 10.1038/nclimate1388

Hong, Y. W., D. X. Yuan, Q. M. Lin and T. L. Yang. 2008. Accumulation and biodegradation of phenanthrene and fluoranthene by the algae enriched from a mangrove aquatic ecosystem. *Marine Pollution Bulletin*, 56(8): 1400–1405. 10.1016/j.marpolbul.2008.05.003

Jamali, A. A., F. Akbari, M. M. Ghorakhlu, M. de la Guardia and A. Y. Khosroushahi. 2012. Applications of Diatom as Potential Microalgae in Nanobiotechnology. *BioImpacts*, 2(2): 83–89. 10.5681/bi.2012.012

Jauffrais, T., B. Jesus, E. Metzger, J.-L. Mouget, F. Jorissen and E. Geslin. 2016. Effect of light on photosynthetic efficiency of sequestered chloroplasts in intertidal benthic foraminifera (*Haynesina germanica* and *Ammonia tepida*). *Biogeosciences*, 13(9): 2715–2726. 10.5194/bg-13-2715-2016

Jesus, B., V. Brotas, L. Ribeiro, C. R. Mendes, P. Cartaxana and D. M. Paterson. 2009. Adaptations of microphytobenthos assemblages to sediment type and tidal position. *Continental Shelf Research*, 29(13): 1624–1634. 10.1016/j.csr.2009.05.006

Kaparapu, J. 2018. Application of microalgae in aquaculture. *Phykos*, 48: 21–26.

Karl, D. M., M. J. Church, J. E. Dore, R. M. Letelier and C. Mahaffey. 2012. Predictable and efficient carbon sequestration in the North Pacific Ocean supported by symbiotic nitrogen fixation. *Proceedings of the National Academy of Sciences of the United States of America*, 109, 1842–1849. 10.1073/pnas.1120312109

Karsten, U., K. Kuriyama, T. Hübener and J. Woelfel. 2021. Benthic diatoms on sheltered coastal soft bottoms (Baltic Sea)—seasonal community production and respiration. *Journal of Marine Science and Engineering*, 9(9): 949. 10.3390/jmse9090949

Kelly, M. G., J. Penny and B. A. Whitton. 1995. Comparative performance of benthic diatom indices used to assess river water quality. *Hydrobiologia* 302, 169–188. 10.1007/BF00032108

Kerr, B. J., T. A. Kellner and G. C. Shurson. 2015. Characteristics of lipids and their feeding value in swine diets. *Journal of Animal Science and Biotechnology*, 6(1): 1–23. 10.1186/s40104-015-0028-x

Khoo, K. S., W. Y. Chia, K. W. Chew and P. L. Show. 2021. Microalgal-bacterial consortia as future prospect in wastewater bioremediation, environmental management and bioenergy production. *Indian Journal of Microbiology*, 61: 262–269. 10.1007/s12088-021-00924-8

Kim, M-C., J-H. Kim and G-T. Seo. 2014. Removal Improvement in Water Treatment Plant for Occurrence of Diatoms (*Synedra* sp.) in the Nakdong River, *Journal of Korean Society of Environmental Engineers*, 36 (1): 29–34. 10.4491/KSEE.2014.36.1.29

Kolkwitz, R., and M. Marsson. 1908. Ökologie der pflanzlichen Saprobien. *Berichte der Deutschen Botanischen Gesellschaft*, 26a: 505–519.

Kooistra, W. H., R. Gersonde, L. K. Medlin, and D. G. Mann. 2007. The origin and evolution of the diatoms: their adaptation to a planktonic existence. In *Evolution of primary producers in the sea*, ed. P. Falkowski, and A. H. Knoll, 207–249. London: Academic Press.

Koyande, A. K., K. W. Chew, K. Rambabu, Y. Tao, D. T. Chu, and P. L. Show. 2019. Microalgae: A potential alternative to health supplementation for humans. *Food Science and Human Wellness*, 8(1): 16–24. 10.1016/j.fshw.2019.03.001

Lavaud, J., R. F. R. Strzepek and P. P. G. Kroth. 2007. Photoprotection capacity differs among diatoms: possible consequences on the spatial distribution of diatoms related to fluctuations in the underwater light climate. *Limnology and Oceanography*, 52(3): 1188–1194. 10.4319/lo.2007.52.3.1188

Li, M. H., E. H. Robinson, C. S. Tucker, B. B. Manning and L. Khoo. 2009. Effects of dried algae *Schizochytrium* sp., a rich source of docosahexaenoic acid, on growth, fatty acid composition, and sensory quality of channel catfish *Ictalurus punctatus*. *Aquaculture*, 292(3), 232–236. 10.1016/j.aquaculture.2009.04.

Litchman, E., Klausmeier, C. A., and Bossard, P. 2004. Phytoplankton nutrient competition under dynamic light regimes. *Limnology and Oceanography*, 49(4 II): 1457–1462. 10.4319/lo.2004.49.4_part_2.1457

Lu, X., Y. Liu, and Y. Fan. 2020. Diatom taxonomic composition as a biological indicator of the ecological health and status of a river basin under agricultural influence. *Water*, 12: 2067. 10.3390/w12072067

Loix, C., M. Huybrechts, J. Vangronsveld, M. Gielen, E. Keunen and A. Cuypers. 2017. Reciprocal Interactions between Cadmium-Induced Cell Wall Responses and Oxidative Stress in Plants. *Frontiers in Plant Science*, 8: 1867. 10.3389/fpls.2017.01867

López-Pedrouso, M., J. M. Lorenzo, J. Cantalapiedra, C. Zapata, J. M. Franco and D. Franco. 2020. Aquaculture and by-products: Challenges and opportunities in the use of alternative protein sources and bioactive compounds. In *Advances in Food and Nutrition Research*, ed. J. M. Lorenzo, and F. J. Barba, 92, 127–185. London: Academic Press.

Malinowska-Gniewosz, A., J. Czerwik-Marcinkowska, A. Massalski, A. Kubala-Kukus, U. Majewska and M. Jankowski. 2018. Relationships between diatoms and environmental variables in industrial water biotopes of Trzuskawica S.A. (Poland). *Open Chemistry*, 16: 272–282. 10.1515/chem-2018-0033

Malviya, S., E. Scalco, S. Audic, F. Vincent, A. Veluchamy, J. Poulain, P. Wincker, D. Iudicone, C. De Vargas, L. Bittner, A. Zingone and C. Bowler. 2016. Insights into global diatom distribution and diversity in the world's ocean. *Proceedings of the National Academy of Sciences of the United States of America*, 113(11): E1516–E1525. 10.1073/pnas.1509523113

Manan, H. and M. Ikhwanuddin. 2021. Triploid induction in penaeid shrimps aquaculture: A review. *Reviews in Aquaculture*, 13(1): 619–631. 10.1111/raq.12489

Marchetti, A. and N. Cassar. 2009. Diatom elemental and morphological changes in response to iron limitation: a brief review with potential paleoceanographic applications. *Geobiology*, 7(4): 419–431. 10.1111/j.1472-4669.2009.00207.x

Marella, T. K., I. Y. Lopez-Pacheco, R. Parra-Saldivar, S. Dixit and A. Tiwari. 2020. Wealth from waste: Diatoms as tools for phycoremediation of wastewater and for obtaining value from the biomass. *Science of the Total Environment*, 724: 1–17. 10.1016/j.scitotenv.2020.137960

Masmoudi, S., N. Nguyen-Deroche, A. Caruso, H. Ayadi, A. Morant-Manceau, G. Tremblin, M. Bertrand, and B. Schoefs. 2013. Cadmium, copper, sodium and zinc effects on diatoms: from heaven to hell – A review. *Cryptogamie, Algologie*, 34(2): 185–225. 10.7872/crya.v34.iss2.2013.185

Mau, A. and R. Jha. 2018. Aquaculture of two commercially important molluscs (abalone and limpet): existing knowledge and future prospects. *Reviews in Aquaculture*, 10(3): 611–625. 10.1111/raq.12190

McMinn, A., A. Martin and K. Ryan. 2010. Phytoplankton and sea ice algal biomass and physiology during the transition between winter and spring (McMurdo Sound, Antarctica). *Polar Biology*, 33(11): 1547–1556. 10.1007/s00300-010-0844-6

Medley, C. N. and W. H. Clements. 1998. Responses of diatom communities to heavy metals in streams: The influence of longitudinal variation. *Ecological Applications*, 8: 631–644. 10.1890/1051-0761(1998)008[0631:RODCTH]2.0.CO;2

Migné, A., Spilmont, N., Boucher, G., Denis, L., Hubas, C., Janquin, M.-A., Rauch, M., and Davoult, D. (2009). Annual budget of benthic production in Mont Saint-Michel Bay considering cloudiness, microphytobenthos migration, and variability of respiration rates with tidal conditions. *Continental Shelf Research*, 29(19),2280–2285. 10.1016/j.csr.2009.09.004.

Molina-Cárdenas, C. A., and M. del Pilar Sánchez-Saavedra. 2017. Inhibitory effect of benthic diatom species on three aquaculture pathogenic *Vibrios*. *Algal Research*, 27: 131–139. 10.1016/j.algal.2017.09.004

Monroig, Ó., A. C. Shu-Chien, N. Kabeya, D. R. Tocher and L. F. C. Castro. 2022. Desaturases and elongases involved in long-chain polyunsaturated fatty acid biosynthesis in aquatic animals: From genes to functions. *Progress in Lipid Research*, 101157. 10.1016/j.plipres.2022.101157

Morin, S., N. Corcoll, B. Bonet, A. Tlili, and H. Guasch. 2014. Diatom responses to zinc contamination along a Mediterranean river. *Plant Ecology and Evolution*, 147: 325–332. 10.5091/plecevo.2014.986

Naviner, M., J. P. Bergé, P. Durand and H. Le Bris. 1999. Antibacterial activity of the marine diatom *Skeletonema costatum* against aquacultural pathogens. *Aquaculture*, 174(1–2), 15–24. 10.1016/S0044-8486(98)00513-4

Nazmi, A., R. Hauck, A. Davis, M. Hildebrand, L. B. Corbeil and R. A. Gallardo. 2017. Diatoms and diatomaceous earth as novel poultry vaccine adjuvants. *Poultry science*, 96(2), 288–294. 10.3382/ps/pew250

Noh, Y. H., K. H. Kim, H. N. Moon, G. M. Go and I. K. Yeo. 2020. Effect of Dietary Supplementation of Diatom *Melosira nummuloides* and Lactic Acid Bacteria *Lactobacillus plantarum* on the Growth and Immune Stimulation Responses of Olive Flounder *Paralichthys olivaceus*. *Korean Journal of Fisheries and Aquatic Sciences*, 53(4): 597–605. 10.5657/KFAS.2020.0597

Norambuena, F., K. Hermon, V., Skrzypczyk, J. A. Emery, Y. Sharon, A. Beard and G. M. Turchini. 2015. Algae in fish feed: performances and fatty acid metabolism in juvenile Atlantic salmon. *PloS one*, 10(4): e0124042. 10.1371/journal.pone.0124042

Oruganti, R. K., K. Katam, P. L. Show, V. Gadhamshetty, V. K. K. Upadhyayula and D. Bhattacharyya. 2022. A comprehensive review on the use of algal-bacterial systems for wastewater treatment with emphasis on nutrient and micropollutant removal. *Bioengineered*, 13(4): 10412–10453. 10.1080/21655979.2022.2056823

Pajares, S. and R. Ramos. 2019. Processes and microorganisms involved in the marine nitrogen cycle: Knowledge and gaps. *Frontiers in Marine Science*, 6: 739. 10.3389/fmars.2019.00739

Palmer, C. M. 1969. A composite rating of algae tolerating organic pollution. *Journal of Phycology*, 5(1): 78–82. 10.1111/j.1529-8817.1969.tb02581.x

Peter, A. P., K. S. Khoo, K. W. Chew, T. C. Ling, S-H. Ho, S-J. Chang and P. L. Show. 2021. Microalgae for biofuels, wastewater treatment and environmental monitoring. *Environmental Chemistry Letters*, 19(4): 2891–2904. 10.1007/s10311-021-01219-6

Pniewski, F. and I. Piasecka-Jędrzejak. 2020. Photoacclimation to Constant and Changing Light Conditions in a Benthic Diatom. *Frontiers in Marine Science*, 7(5): 1–13. 10.3389/fmars.2020.00381

Pocha, C. K. R., W. Y. Chia, K. W. Chew, H. S. H. Munawaroh, and P. L. Show. 2022. Current advances in recovery and biorefinery of fucoxanthin from *Phaeodactylum tricornutum*. *Algal Research*, 65: 102735. 10.1016/j.algal.2022.102735

Poikane, S., G. Varbiro, M. G. Kelly, S. Birk and G. Phillips. 2021. Estimating river nutrient concentrations consistent with good ecological condition: More stringent nutrient thresholds needed. *Ecological Indicators*, 121: 107017. 10.1016/j.ecolind.2020.107017

Pyle, A. E., A. M. Johnson and T. A. Villareal. 2020. Isolation, growth, and nitrogen fixation rates of the *Hemiaulus-Richelia* (diatom-cyanobacterium) symbiosis in culture. *PeerJ*, 8: e10115. 10.7717/peerj.10115

Raji, A. A., P. A. Alaba, H. Yusuf, N. H. Abu Bakar, N. M. Taufek, H. Muin, Z. Alias, P. Milow and S. Abdul Razak. 2018. Fishmeal replacement with *Spirulina platensis* and *Chlorella vulgaris* in African catfish (*Clarias gariepinus*) diet: Effect on antioxidant enzyme activities and haematological parameters. *Research in Veterinary Science*, 119: 67–75. 10.1016/j.rvsc.2018.05.013

Ragg, N. L., N. King, E. Watts and J. Morrish. 2010. Optimising the delivery of the key dietary diatom *Chaetoceros calcitrans* to intensively cultured Greenshell™ mussel larvae, *Perna canaliculus*. *Aquaculture*, 306(1–4): 270–280. 10.1016/j.aquaculture.2010.05.010

Raven, J. A. and J. Beardall. 2021. Influence of global environmental change on plankton. *Journal of Plankton Research*, 43(6): 779–800. 10.1093/plankt/fbab075

Renuka, N., A. Sood, R. Prasanna and A. Ahluwalia. 2015. Phycoremediation of wastewaters: A synergistic approach using microalgae for bioremediation and biomass generation. *International Journal of Environmental Science and Technology*, 12(4): 1443–1460. 10.1007/s13762-014-0700-2

Rost, B., Zondervan, I. and D. Wolf-Gladrow. 2008. Sensitivity of phytoplankton to future changes in ocean carbonate chemistry: Current knowledge, contradictions and research directions. *Marine Ecology Progress Series*, 373: 227–237. 10.3354/meps07776

Round, F., R. Crawford and D. Mann. 1990. *Diatoms: biology and morphology of the genera*. Cambridge: Cambridge University Press.

Rynearson T. A., K. Richardson, R. S. Lampitt, M. E. Sieracki, A. J. Poulton, M. M. Lyngsgaard and M. J. Perry. 2013. Major contribution of diatom resting spores to vertical flux in the sub-polar North Atlantic. *Deep Sea Research Part I: Oceanographic Research Papers*, 82: 60–71. 10.1016/j.dsr.2013.07.013

Saleh, N. E., R. F. Ismail, A. E. D. H. Sayed, E. H. Zaghloul and H. Saleh. 2020. Comprehensive assessment of benthic diatom (*Amphora coffeaeformis*) as a feed additive in Nile tilapia (*Oreochromis niloticus*) diet. *Aquaculture Research*, 51(9): 3506–3519. 10.1111/are.14686

Salleh, S. and A. McMinn. 2021. Response of tropical marine benthic diatoms exposed to elevated irradiance and temperature. *Biogeosciences*, 2021: 1–25. 10.5194/bg-2021-18

Sabine, C. L., R. A. Feely, N. Gruber, R. M. Key, K. Lee, J. L. Bullister, R. Wanninkhof, C. S. L. Wong, D. W. Wallace, B. Tilbrook and F. J. Millero. 2004. The oceanic sink for anthropogenic CO_2. *Science*, 305(5682): 367–371. 10.1126/science.109740

Savelli, R., J. Serôdio, P. Cugier, V. Méléder, P. Polsenaere, C. Dupuy and V. Le Fouest. 2021. Potential Impact of Photoinhibition on Microphytobenthic Primary Production on a Large Intertidal Mudflat. *Journal of Geophysical Research: Biogeosciences*, 126(9): e2021JG006443. 10.1029/2021JG006443

Schvarcz, C. R., S. T. Wilson, M. Caffin, R. Stancheva, Q. Li, K. A. Turk-Kubo, A. E. White, D. M. Karl, J. P. Zehr and G. F. Steward. 2022. Overlooked and widespread pennate diatom-diazotroph symbioses in the sea. *Nature Communications*, 13(1): 799. 10.1038/s41467-022-28065-6

Sharma, N., D. P. Simon, A. M. Diaz-Garza, E. Fantino, A. Messaabi, F. Meddeb-Mouelhi, H. Germain and I. Desgagné-Penix. 2021. Diatoms biotechnology: various industrial applications for a greener tomorrow. *Frontiers in Marine Science*, 8: 636613. 10.3389/fmars.2021.636613

Shi, D., S. A. Kranz, J.-M. Kim, and F. M. M. Morel. 2012. Ocean acidification slows nitrogen fixation and growth in the dominant diazotroph *Trichodesmium* under low-iron conditions. *Proceedings of the National Academy of Sciences*, 109(45): E3094–E3100. 10.1073/pnas.1216012109

Show, P. L., M. S. Y. Tang, D. Nagarajan, T. C. Ling, C-W. Ooi and J-S. Chang. 2017. A holistic approach to managing microalgae for biofuel application. *International Journal of Molecular Sciences*, 18(1): 215. 10.3390/ijms18010215

Show, P. L., S. Thangalazhy-Gopakumar and D. C. Y. Foo. 2021. Sustainable technologies for waste reduction and pollutants removals. *Clean Technologies and Environmental Policy*, 23(1): 1–2. 10.1007/s10098-020-02001-y

Shukla, S. K., R. Mohan and M. Sudhakar. 2009. Diatoms: a potential tool to understand past oceanographic settings. *Current Science*, 97(12): 1726–1734. https://www.jstor.org/stable/24107252

Snoeijs, P., S. Busse and M. Potapova. 2002. The importance of diatom cell size in community analysis. *Journal of Phycology*, 38(2): 265–281. 10.1046/j.1529-8817.2002.01105.x

Sørensen, M., G. M. Berge, K. I. Reitan and B. Ruyter. 2016. Microalga *Phaeodactylum tricornutum* in feed for Atlantic salmon (*Salmo salar*)—Effect on nutrient digestibility, growth and utilization of feed. *Aquaculture*, 460: 116–123. 10.1016/j.aquaculture.2016.04.010

Sorgeloos, P., P. Dhert and P. Candreva. 2001. Use of the brine shrimp, *Artemia* spp., in marine fish larviculture. *Aquaculture*, 200(1–2): 147–159. 10.1016/S0044-8486(01)00698-6

Syafaat, M. N., M. N. Azra, K. Waiho, H. Fazhan, A. B. Abol-Munafi, S. D. Ishak, M. Syahnon, A. Ghazali, H. Ma and M. Ikhwanuddin. 2021. A review of the nursery culture of mud crabs, genus *Scylla*: current progress and future directions. *Animals*, 11(7): 2034. 10.3390/ani11072034

Tanaka, T., K. Yoneda and Y. Maeda. 2022. Lipid Metabolism in Diatoms. In *The Molecular Life of Diatoms*, ed. A. Falciatore, and T. Mock, 493–527. Switzerland: Springer, Cham.

Tiwari, A. and T. K., Marella. 2019. Potential and application of diatoms for industry-specific wastewater treatment. In *Application of Microalgae in Wastewater treatment*, ed. S. K. Gupta and F. Bux, 321–339. Switzerland: Springer, Cham.

Trushenski, J. T. and R. T. Lochmann. 2009. Potential, Implications and Solutions Regarding the Use of Rendered Animal Fats in Aquafeeds. *American Journal of Animal and Veterinary Sciences*, 4(4): 108–128.

Uriarte, I., M. Astorga, J. C. Navarro, M. T. Viana, C. Rosas, C. Molinet, J. Hernández, J. Navarro, I. Moreno-Villoslada, R. Amthauer and G. Kausel. 2019. Early life stage bottlenecks of carnivorous molluscs under captivity: a challenge for their farming and contribution to seafood production. *Reviews in Aquaculture*, 11(3): 431–457. 10.1111/raq.12240

Valenzuela, J. J., A. López García de Lomana, A. Lee, E. V. Armbrust, M. V. Orellana and N. S. Baliga. 2018. Ocean acidification conditions increase resilience of marine diatoms. *Nature Communications*, 9(1): 2328. 10.1038/s41467-018-04742-3

Vieira, S., L. Ribeiro, B. Jesus, P. Cartaxana, and J. M. da. Silva. 2013. Photosynthesis Assessment in Microphytobenthos Using Conventional and Imaging Pulse Amplitude Modulation Fluorometry. *Photochemistry and Photobiology*, 89(1): 97–102. 10.1111/j.1751-1097.2012.01224.x

Villareal, T. A. 1991. Nitrogen-fixation by the cyanobacterial symbiont of the diatom genus *Hemiaulus*. *Marine Ecology Progress Series*, 76, 201–204.

Voss, M., H. W. Bange, J. W. Dippner, J. J. Middelburg, J. P. Montoya and B. Ward. 2013. The marine nitrogen cycle: recent discoveries, uncertainties and the potential relevance of climate change. *Philosophical Transactions of the Royal Society B – Biological Sciences*, 368(1621): 21030121. 10.1098/rstb.2013.0121

Volland, S., E. Bayer, V. Baumgartner, A. Andosch, C. Lutz, E. Sima, U. Lutz-Meindl. 2014. Rescue of heavy metal effects on cell physiology of the algal model system *Micrasterias* by divalent ions. *Journal of Plant Physiology*, 171(2): 154–163. 10.1016/j.jplph.2013.10.002

Wu, Y., Z. Li, W. Du and K. Gao. 2015. Physiological response of marine centric diatoms to ultraviolet radiation, with special reference to cell size. *Journal of Photochemistry and Photobiology B: Biology*, 153: 1–6. 10.1016/j.jphotobiol.2015.08.035

Yong, J. J. Y., K. W. Chew, K. S. Khoo, P. L. Show and J-S. Chang. 2021. Prospect and development of algal-bacterial biotechnology in environmental management and protection. *Biotechnology Advance*, 47: 107784. 10.1016/j.biotechadv.2020.107684

Yuan, C., Z. Xu, X. Zhang, Q. Wei, H. Wang and Z. Wang. 2019. Photosynthetic physiologies of phytoplankton in the eastern equatorial Indian Ocean during the spring inter-monsoon. *Acta Oceanologica Sinica*, 2016, 1–9. 10.1007/s13131-018-1218-0

Xi, H., Losa, S. N., Mangin, A., Soppa, M.a A., Garnesson, P., Demaria, J., Liu, Y., d'Andon, O. H. F., and Bracher, A. (2020). Global retrieval of phytoplankton functional types based on empirical orthogonal functions using CMEMS GlobColour merged products and further extension to OLCI data. *Remote Sensing of Environment*, 240, 11170410.1016/j.rse.2020.111704.

Xing, R. L., C. H. Wang, X. B. Cao and Y. Q. Chang. 2007. The potential value of different species of benthic diatoms as food for newly metamorphosed sea urchin *Strongylocentrotus intermedius*. *Aquaculture*, 263: 142–149. 10.1016/j.aquaculture.2006.10.003

Xing, R. L., Z. M. Qian, C. H. Wang and N. Tang. 2008. Isolation of benthic marine diatom species and their morphologies and nutritional component. *Journal of Harbin Institute of Technology*, 40(9): 1495–1500.

Xing R-L., W-W. Ma, Y-W. Shao, X-B. Cao, C. Su, H-W. Song, Q. Su and G-F. Zhou. 2018. Growth and potential purification ability of *Nitzschia* sp. benthic diatoms in sea cucumber aquaculture wastewater. *Aquaculture Research*, 49(8): 2644–2652. 10.1111/are.13722

7 Climate Change and Diatoms

Lalit Kumar Pandey
Department of Plant Science, Faculty of Applied Sciences, MJP Rohilkhand University, Bareilly

E.A. Bergey
Oklahoma Biological Survey and Department of Biology, University of Oklahoma, Norman, OK, USA

Taejun Han
Bio Environmental Science and Technology (BEST) Lab, Ghent University Global Campus, Songdomunhwa-ro, Incheon, Korea

Department of Animal Sciences and Aquatic Ecology, Ghent University, Ghent, Belgium

Jihae Park
Bio Environmental Science and Technology (BEST) Lab, Ghent University Global Campus, Songdomunhwa-ro, Incheon, Korea

Center for Environmental and Energy Research, Ghent University Global Campus, Songdomunhwa-ro, Incheon, Korea

Department of Animal Sciences and Aquatic Ecology, Ghent University, Ghent, Belgium

Sudeeksha Negi, Mahima Gupta, Divyanshi Gupta, Abhishek Kumar, Shivangi Singh, and Sangeeta
Department of Plant Science, Faculty of Applied Sciences, MJP Rohilkhand University, Bareilly

Sanjeev Kumar Maurya
Department of Plant Science, Faculty of Applied Sciences, MJP Rohilkhand University, Bareilly

Constituent Government College, Puranpur, Pilibhit

CONTENTS

DOI: 10.1201/9781003322115-7

7.1 INTRODUCTION

Climate change is a global problem that has been ongoing for at least four decades. Although numerous efforts are being made around the world to address and solve this problem, significant concrete steps are still needed. Climate change is an encompassing term that includes a spectrum of catastrophic events in ecosystems, including aquatic ecosystems. Aquatic ecosystems are affected by warming (IPCC, 2018), acidification (Sunday et al., 2014), and deoxygenation (Limburg et al., 2020) (i.e., Trio effect), processes that lead to sea-level rise (Fleming et al., 2018), formation of harmful algal blooms (IPCC, 2018), unprecedented loss of biodiversity (Cavicchioli et al., 2019), extreme weather events (Vose et al., 2017), and human health hazards in the form of increasing vector and waterborne disease (e.g., malaria, dengue fever, Zika virus disease, cholera) (Vose et al., 2017; Cavicchioli et al., 2019). In addition, climate change may worsen air, water, and soil conditions, leading to unpreceded increases in floods and droughts, deterioration of coastal ecosystems, and disruption of animal and plant life-cycles (IPCC, 2018). Despite these impacts, powerful developed nations are reluctant to fully accept the phenomenon of climate change and its multiple causes, which increases the difficulty of implementing significant solutions. A major reason for the denial is perceived short-term financial prospects. If governments accept the reality of climate change, they will have to reduce significantly carbon emissions (especially of carbon dioxide), which may adversely affect their economies. Another reason behind their denial is the difficulty of proving that human-caused climate change produces the observed changes. Climate change is a slow process that began with the origin of human technological advancements and has accelerated in pace in the last few decades. Evidence of climate change as the causative factor are indicated by temporal physio-chemical changes, such as the increasing level of carbon dioxide in the atmosphere compared to background levels (i.e., 410 ppm CO_2), rises in sea level, acidification, etc. In contrast to physio-chemical changes, our understanding of the effects of climate change on living organisms is less well known, and research in this area is urgently needed.

At least three biological effects of climate change on living organisms have been reported: first, changes in phenology (Durant et al., 2007); second, changes in species distributions (Visser and Both, 2005); and third, changes in the nutrient content of autotrophs. Reduction in size may be an additional effect of climate change on living organisms (Daufresne et al., 2009). Size reduction under climate change has been observed in fish and birds.

Although the effects of climate change on larger organisms are regularly reported, the effects on microorganisms are less well known. In this chapter, we attempt to decipher the effects of climate change on primary producers, especially diatoms, and to understand how the phenomenon of climate change affects the various morphological and physiological aspects of diatom communities in aquatic ecosystems, including both marine and freshwater ecosystems.

7.2 COMMUNITY SHIFTS

Aquatic ecosystems are continuously being altered by the process of climate change. In the coming decades, lakes and reservoirs may experience progressive warming, acidification, and deoxygenation, which will affect various human-ecosystem interactions, such as aquaculture and fisheries, tourism and recreational use, and public health (Gobler, 2020). Community shifts in primary producers is a prominent effect of climate change (Cavicchioli et al., 2019). Oceanic ecosystems have especially high phytoplanktonic biodiversity, and each taxon is adapted to specific water conditions. Changes in the physio-chemical properties of the water, such as nutrient content, salinity, turbidity, etc., alter the species composition of inhabiting phytoplankton in that particular environment.

Many plankton taxa move vertically in the water column to replenish nutrients. Ocean warming, which began in the 1950s, has increased thermal stratification and disrupted normal nutrient cycling. Greater accumulation of carbon dioxide and continued warming of seawater is increasing the abundance of smaller diatom species relative to larger ones. This size-dependent shift in planktonic diatom species has been documented in various locations around the world, but how this shift alters the aquatic food web and overall aquatic productivity remains largely unexplored (Bopp, 2005). Xiao et al. (2018) studied the diversity of diatoms and dinoflagellates in relation to warming and eutrophication in the East China Sea by analyzing a large number of samples collected between 2002 and 2015 to model the effects of environmental changes on these two taxa. A major result of this study is that the combined effects of warming and eutrophication leads to a higher abundance of dinoflagellates relative to diatoms, along with changes in the pigment composition (fucoxanthin and peridinin) of the primary producers. In the same study, they also predicted that if the current trajectory continues until 2100, diatom populations will decrease by 19% and dinoflagellate populations will increase by 60% in the Eastern China Sea, which will likely increase the frequency and severity of harmful dinoflagellate blooms. Similarly, Lazarus et al. (2014) studied the effects of climate change on Cenozoic planktonic marine diatom diversity by using the Neptune Sandbox Berlin NSB dataset hosted at the Museum für Naturkunde in Berlin. This study showed that estimated past atmospheric pCO_2 and the global carbon isotope record are strongly correlated with 15 million years of diatom diversity. Furthermore, the researchers speculated that if oceanic warming continues at the current pace, some planktonic diatom species may be threatened with extinction in the near future, which would negatively impact the carbon sequestration and biological pump of oceanic ecosystems.

Similarly, Jin and Agustí (2018) found that ocean warming due to climate change resulted in changes to the phenology of marine inhabitants and demonstrated poleward shifts in distributions. Information regarding adaptive strategies adopted by phytoplankton to combat warming seas is incomplete. In order to avoid possible biotic desertification and reduction in their biodiversity, local plankton species will need to undergo phenotypic and even evolutionary adaptation to sustain themselves in the warmer water. In the oceanic cold waters and also in the tropical seas, diatom microalgae are the chief primary producers, playing a crucial role in maintaining the biological pump of nutrients to biomass. Jin and Agustí (2018) carried out long-term experiments on four isolated diatom taxa (*Chaetoceros tenuissimus*, *Chaetoceros* sp., *Thalassiosira* sp., and *Synedra* sp.) from the tropical Red Sea, allowing 200–600 generations to adapt to warming conditions (30 °C) under different thermal strategies. The diatom taxa *Chaetoceros tenuissimus* and *Synedra* sp. utilized a “hotter is better” strategy by shifting their optimum temperature (T_{opt}) and increasing their growth (μ_{max}) under warming conditions. In contrast, warming adapted cells of *Chaetoceros* sp. and *Thalassiosira* sp. did not show a shift in their optimum temperature but did show a shift in their critical thermal minimum (CT_{min}) and maximum (CT_{max}). Under warming conditions, tropical diatoms adapted at the expense of their photosynthetic efficiency and had a lower growth rate, which may ultimately lead to compromised competitive fitness. Through these adaptive strategies, phytoplanktonic diatoms of tropical waters may arrest a possible decline in their diversity and maintain their normal abundance under changing climatic scenario.

Furthermore, Sienkiewicz et al. (2017) studied environmental change data over the last 3,100 years based on a sediment core from the Arctic Revvatnet Lake (Hornsund area, WS Spitsbergen, Svalbard). This study showed that until the middle of the 19th century (including the little ice age), good growth conditions led to the dominance of the planktonic diatom *Cyclotella*. During this period, small and less silicified *Cyclotella sensu lato* developed under a fairly stable, stratified water column that had periodic mixing. Since the beginning of the 20th century, climate warming has increased the sedimentation rate in the catchment area of the lake due to increased erosion. The pattern of thermal stratification of the water column was impacted, nutrient concentrations increased, and ultimately the diatom species composition was affected, causing benthic forms to dominate while almost all planktonic forms either decreased in abundance or disappeared. Such a change would affect the lake food web, as filter feeders, especially zooplankton, are impacted.

7.3 BLOOM FORMATION

Since the 1980s, both climatic and non-climatic factors, such as increased nutrient run-off resulting from human activities, have led to an expansion in the distribution and an increase in the frequency of algal blooms in aquatic ecosystems. Factors contributing to algal blooms include ocean and lake warming, acidification, oxygen depletion, heat waves, eutrophication, and pollution. Algal blooms can have adverse impacts on the economy, tourism, food availability, human health, and native ecosystems. Climate change is leading to higher air temperatures, which may have a corresponding effect on melting glacier ice in polar regions. Higher water temperature and melting (freshwater) ice change the chemical stoichiometry of oceanic water, influencing their inhabitant flora, especially algae. Since oceanic waters are dominated by phytoplanktonic diatoms, more research on diatom abundance in these altered environments is needed. These physical and chemical changes in the aquatic environment may result in more frequent algal blooms, which are caused mainly by diatoms and dinoflagellates.

The potential of climate change to affect diatom blooms was demonstrated in a mesocosm experiment with a phytoplankton community from Kielfjord, Germany, to investigate the effects of oceanic warming and acidification on biogeochemical element cycling and species composition (Sett et al., 2018). In all treatments, a diatom-dominated (*Skeletonema costatum*) bloom developed, along with the increase in their biomass and the abundance of larger diatom species (size greater than 20 μm, with taxa such as *Thalassionema nitzschioides*, *Thalassiosira* sp., *Nitzschia longissimi*, *Navicula transitans*, *Fragilaria* sp., *Guinardia* sp., and *Chaetoceros* sp.).

The micronutrient iron plays an important role in the primary productivity and the overall biogeochemical cycle of the ocean. For example, Lampe et al. (2018) investigated the role of iron addition and removal on natural phytoplankton communities from varying iron environments of coastal and oceanic sites of the California upwelling zone. Bloom-forming diatom genera (i.e., *Pseudo-nitzschia, Chaetoceros,* and *Thalassiosira*) adopted different mechanisms to cope with iron limited and enriched conditions. Ferritin-dependent iron removal and long-term storage favor *Pseudo-nitzschia,* whereas vacuolar storage was associated with *Chaetoceros* and *Thalassiosira.* Under different iron conditions (low and high) as well as under iron-limitation, *Pseudo-nitzschia* demonstrated high iron storage capabilities.

Water temperature, alone, may not be sufficient to trigger diatom blooms. Costa et al. (2021) examined the dynamics of an extensive phytoplanktonic diatom bloom (in terms of composition and biomass) near the Northern Antarctic Peninsula, in relation to the stability of the upper mixed layer of the water column. The large centric diatom *Odontella weissflogii* (mostly > 70 μm in length) dominated the diatom bloom, which was associated with a high uptake of CO_2 (-60 mmol m^{-2} d^{-1}) in localized areas. The authors hypothesized that water column stability was the cause of the intense diatom bloom in the Gerlache Strait. Similarly, Costa et al. (2021) studied the effect of the 2015-2016 El Niño in the Antarctic Peninsula area during cool conditions and reported a bloom of large diatoms on the basis of the phytoplankton in collected water samples and satellite chlorophyll *a* data. They hypothesized that rising oceanic and atmospheric temperatures led to increased melting of polar ice, which ultimately resulted in episodic diatom blooms. Likewise, temperature alone did not produce the massive diatom bloom in the Lake Berleber (central Germany) in winter 2019 (Kong et al., 2020), where intense monitoring over a one-year period indicated that winter blooms were caused by a combination of light and temperature, whereas temperature alone had little effect.

Unlike most non-diatom algal blooms, diatom blooms generally occur in cool or cold waters (e.g., polar seas or below the surface of lakes), areas where blooms may be less obvious and that may be less monitored. Understanding the causes of diatom blooms requires better knowledge of the interactions among climate change effects, various biogeochemical cycles, and ecological dynamics of algae (especially diatoms).

7.4 SIZE REDUCTION

Two overall effects of climate change have been well documented; these are changes in the distribution patterns of organisms and changes in their phenology. More recently, size reduction of some organisms has been recognized as a potential effect of climate change. Size reduction has been documented in birds and fishes. In contrast, the effect of climate change on algal size is unclear. In this regard, diatoms may show size reduction under climate change conditions (Figure 7.1). The main reason for size reduction in diatoms is dilution of ocean water, which results in the dilution of the major and minor nutrients found in the water. Major nutrients, especially nitrate and phosphate, as well as minor nutrients, such as the metals Cu and Fe, play important roles in the growth of diatoms. Under diluted-water conditions, the concentrations of limiting nutrients required for diatom growth are reduced, which can result in reduced population size or reduced individual size in order to compensate for lower nutrient conditions.

Body size of an organism is an important ecological feature, which may show plasticity in response to alterations in the environment, including anthropogenic disturbances (Abonyi et al., 2020). Alteration in body size affects the growth and reproduction of an organism and can ultimately affect fitness (Hooper et al., 2005). Therefore, body size is a promising parameter to assess the effect of climate change on some living organisms. Global warming leads to a cascade of events affecting air, water, and soil, which may alter body size through changes in the metabolic

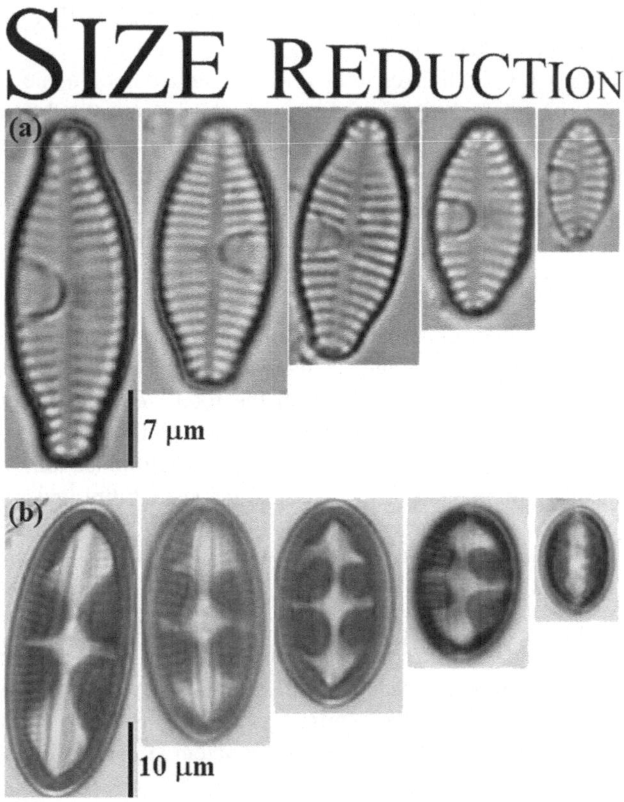

FIGURE 7.1 Dead frustules of diatoms showing size reduction in *Planothidium* sp. (a) and *Diploneis* sp. (b).

rates of organisms (Gillooly et al., 2001). Many studies have demonstrated that alteration in body size due to climate change is a common ecological response (Daufresne et al., 2009) that is observed more frequently in aquatic ecosystems than terrestrial ecosystems (Forster et al., 2012). Aquatic ectothermic organisms, such as phytoplankton, are regularly studied to assess the long-term changes in their cell size (Finkel et al., 2009a & b; Mousing et al., 2014; Smol et al., 2005).

As average air temperatures increase due to global climate change, oceanic and freshwater environments are affected in various ways that in turn affect the residing aquatic biota (IPCC, 2007). Rising temperatures in aquatic water bodies can enhance thermal stratification, and prolonged stratification can produce an upper stratum with higher temperatures and greatly reduced nutrient concentrations (Schmittner, 2005; Winder et al., 2009). Stratification resulting in reduced nutrient availability will favor smaller phytoplankton because of their efficient nutrient uptake capability (Finkel et al., 2009a & b; Irwin and Oliver, 2009; Lewis, 1976).

Oceanic ecosystems cover 70% of the total earth surface and hence support extensive populations of phytoplankton. Among phytoplankton, diatoms provide 40% of primary productivity of oceanic waters and produce 25% of atmospheric oxygen. The ecological significance of diatoms combined with their natural range of sizes within species mean that diatoms are an excellent model group for understanding the impact of climate change on body size. Svensson et al. (2014) studied benthic diatom community cell size in the Baltic Sea in relation to temperature (with other factors, such as salinity and nutrient availability) with special reference to climate change. They found that small diatom cells were associated with high summer temperatures, which ultimately led to lower average community cell volume. They concluded that higher summer temperatures resulted in increased land-runoff of melting glacier ice, which caused a decrease in mean cell size in the benthic diatom community, affecting the dynamics of energy transfer to organisms at higher trophic levels. In freshwater ecosystems, the effect of climate change on the size diminution of diatoms has also been documented. For example, Bramburger et al. (2017) examined diatom cell size trends from a historical dataset (~1900–2015) of the Laurentian Great Lakes (specifically Lakes Superior and Ontario). In both lakes, they documented a demographic shift during the 20th century toward smaller-sized diatom species. Across the basin, mean cell volume of centric diatoms declined by 5.11 $\mu m^3/y$. Similarly, Winder et al. (2009) studied an extensive dataset from Lake Tahoe and found that climate change resulted in increased stratification of lake water, causing reduction in nitrogen to phosphorus ratios, which favored small-celled diatoms (especially the genus *Cyclotella*), whereas the biovolume of the diatoms remained unchanged. The authors concluded that the diverse group of diatoms having a high surface area to volume ratio (i.e., small-sized species) are well adapted to decreased mixing of lake water and, as a result, smaller phytoplanktonic diatom cells dominate under a warmer climate regime. Winder et al. (2009) concluded that a decrease in average diatom size and an increase in the number of smaller diatom cells reduces their sinking velocity, causing alteration in the food web and carbon cycle that affects energy transfer to higher trophic levels. Wilhelm and Adrian (2008) reported increased phosphate availability due to cumulative effects of climate warming and eutrophication. In the late spring and summer, higher availability of phosphate decreased light availability and the Si:P ratio, which favored smaller diatom cells in the community (Finkel et al., 2009 a & b; Kilham et al., 1996).

Dilution of oceanic water due to inflows from melting glacier ice results in the dilution of oceanic nutrients, including metals. Fe plays an important role in the development of silica frustules and is a limiting nutrient for diatoms in some oceanic regions. Martin (1990) presented justification explaining how glacial/interglacial atmospheric CO_2 concentration led to changes in primary productivity in the Southern Ocean. Specifically, size differentiated limitation of growth rates occurred (Martin and Fitzwater, 1988; Martin et al., 1989), as the abundance of large diatoms increased upon addition of Fe to samples from Fe-deficient oceanic waters, demonstrating Fe limitation.

Martin (1990) hypothesised that Southern Ocean phytoplankton were unable to utilize all of the available surface nutrients, especially phosphate and nitrate, due to iron limitation. During the preindustrial period (Holocene interglacial) and the last glacial maximum (LGM), the combination of high CO_2 levels (280 ppm) and 50 times higher Fe levels from atmospheric dust resulted in a higher growth rate of phytoplankton and hence enhanced sequestration of upwelling nutrients, which ultimately led to a LGM decline of atmospheric carbon dioxide to 200 ppm. Since oceanic phytoplankton is dominated by diatoms, Fe enrichment led to rapid growth of planktonic diatoms (with growth enhanced by vegetative reproduction) and supported large diatoms, but as nutrients became limiting, diminution of diatom cells occurred. In contrast, in freshwater environments, metal enrichment (pollution) is a major factor in size diminution in diatom frustules, especially as not all concentrations of metals are beneficial and concentrations can be high. For example, significant size reduction was observed in adnate (*Achnanthidim minutissimum*), colonial (*Asterionella formosa* and *Tabellaria flocculosa*), and motile (*Brachysira vitrea*) forms of diatoms in Lac Dufault (Québec, Canada) with its metal (Cd, Cu, Fe, Pb, and Zn) polluted sediments (Cattaneo et al., 2004). Similarly, Morin and Coste (2006) examined diatom diversity from the Riou Mort and Riou Viou streams of southwest France in relation to metal (Cd and Zn) pollution. The abundant diatom species *Nitzschia palea* and *Gomphonema parvulum* displayed significant size reduction. In an abandoned mining area (contaminated with metals in surface water and sediments) in Portugal, *Brachysira vitrea* showed >60% size reduction (Luís et al. (2011). In a laboratory experiment, different diatom species (*Nitzschia palea, N. dissipata, N. fonticola,* and *Amphipleura pellucida*) showed significant size reduction under arsenic exposure (Barral-Fraga et al., 2016). One possibility to explain this size reduction is that metal-stressed environments lead to a higher rate of cell division in diatoms, which indirectly resulted in dominance of smaller cells in the community (Morin et al., 2012), as average diatom size decreases with cell divisions. Similarly, the oceanic bloom-forming pennate diatom species, *Pseudo-nitschia granii,* showed higher cell division due to enhanced iron storage by ferritin (Marchetti et al., 2009).

In contrast to the above, Adams et al. (2013) reported that diatoms at different organizational levels (species and community levels) did not follow a temperature-size relationship, a result that has also been reported by several other researchers (Gardner et al., 2011; Rüger and Sommer 2012) because other factors affect diatom size. They argued that any body size shifts at the community level are governed by interspecific effects and that intraspecific size changes played a minor role in studies relating to single-species laboratory experiments. In the same context, Woodward et al. (2010) argued that *in situ* community-level studies are required to understand the varying results for the temperature-size relationships in diatoms. In oligotrophic environments, size change in diatoms might be correlated with nutrient absorption. For example, strong late-summer stratification and a longer ice-free season can result in increased frustule diameter or length, which can reduce their sinking rates (Miklasz and Denny, 2010). Climate warming can increase the duration of the growing season and produce a longer period of water column stratification, resulting in greater exposure to nutrients during the spring and fall overturns and the dominance of larger diatom species in the community (Kilham et al., 1996; Kerrigan et al., 2015).

Several mechanisms have been proposed to explain why ectotherms follow temperature-size relationships. As discussed by Sheridan and Bickford (2011), the metabolic rate of ectotherms is higher at higher temperatures. Growth will be limited, however, if resources (especially nutrients) become limiting. A mechanism to compensate for an increased metabolic rate under resource limitation is for organisms to become smaller (Brown et al., 2004).

Climate change–induced alterations in the physical environment may contribute to size reduction. For example, rising temperature can weaken water column mixing, favoring smaller species of phytoplankton, which sink at a slower rate (Winder and Sommer, 2012). This phenomenon (alteration in stratification of the water column) has been reported in both lakes (Winder et al., 2009) and oceans (Morán et al., 2010). Rising temperature may lead to a series of changes, such as reduced dissolved oxygen (deoxygenation), decreased pH (acidification), and alteration of iron levels, all of which may affect phytoplankton cell size (Forster et al., 2012). Clearly, the effects of climate change on diatom size are complex and involve a multitude of factors.

7.5 LIFE-FORMS AND BIODIVERSITY

Climatic changes affect all organisms to some extent, and these effects can be unexpected and complex. Understanding ecological ramifications associated with climate change is pivotal for understanding how observed effects are caused, and how to manage and reduce undesirable consequences of a changing environment (André et al., 2010). Previous sections in this chapter reviewed changes in community composition, effects on diatom abundance to the point of bloom formation (including blooms of dinoflagellates), and climate change factors that may influence the size of diatom species and communities. This section reviews another aspect of climate change on diatoms – that is, effects on their life-form. Diatom taxa are characterized by their physical growth form (life-form), which may indicate the specific habitat of that taxon. For example, planktonic forms are typically floating in the water column, whereas benthic forms are associated with surfaces. Other life-forms include adnate, stalked, mobile, colonial, and tube-dwelling. The life-form composition of diatom communities can be altered by various environmental and anthropogenic perturbations, including climate change.

An example of environmental effects of diatom life forms is found in rock pools of Hanko and Tammisaari in South Finland. Climate change–driven salinity and wind disturbance can affect the benthic biomass and functional traits of diatoms. Low-profile diatom communities dominated under conditions of increased wind disturbance and lower salinity, demonstrating that small spatial changes in the rock pool environments may lead to shifts in the abundance of communities characterized by mobile, colonial, and high-profile diatoms. Altered salinity and wind disturbance have profound effects on residing diatom communities, which can, in turn, alter the functioning of aquatic ecosystems, especially coastal marine ecosystems (Virta and Teittinen, 2022).

The biodiversity of pristine high-altitude rivers is under considerable threat worldwide due to the shrinking cryosphere as a result of climate change. Data sets from the European Alps showed significant responses of diatoms in relation to decreased glacier cover (Fell et al., 2018). Deglaciation is increasing alpha-diversity and density of diatoms, whereas beta-diversity is declining as streams become more similar. With the decline in glacier cover, composition of the diatom assemblage changed, and these changes were associated with increasing stability of riverbeds along with warmer water temperatures. At the species level, diatoms showed a spectrum of responses, such as reduced dominance of *Eunotia trinacria* (a motile form) in areas with declining glacier cover. In contrast, *Gomphonema calcareum* (a motile form), along with seven taxa, were confined to sites with no glacier cover. Fell et al. (2018) also found that sites having glacier cover ≤26% had 19 (22%) taxa that are classified as threatened, endangered, rare, or decreasing as per the Red List of Algae for Germany. Thus, climate change may benefit these rare diatoms, and Red List classifications of these species may need to be reconsidered in the future (Fell et al., 2018). From this study, it is clear that ongoing deglaciation is altering the composition and diversity of this glacier-fed ecosystem, as the process of deglaciation alters the primary production, biogeochemical cycles, and food web dynamics of the aquatic environment.

Liao et al. (2020) investigated the response of diatom communities to climate change and sedimentary events over the last millennium in Lake Basomtso in the southeast Tibetan Plateau. The changing climatic regime affected the growth rate of unicellular algae, and the longer period of thermal stratification and more stable temperatures resulted in a change from a larger form to a smaller form in the planktonic diatom *Pantocsekiella ocellata*, altered overall diatom community size structure, and increased diatom productivity in the lake. Most of the benthic diatom species in the sediment are from the catchment rivers and the littoral zone, as this lake is deep and is well connected with a river system. The observed increase in benthic diatom abundance in the lake sediment may indicate hydrological changes in the nearby catchment area or may be due to geomorphic slumps and landslides on the steep slopes within the lake, which may be attributed to higher temperatures in this region due to climate change. Although changes due to climate change were evident in the diatom record, such changes were not as clear in the record for terrestrial vegetation.

The evolution of environmental adaptation is a complex process, which stems mainly from unfavorable environmental conditions and abiotic stresses. Since the start of the Industrial Revolution, global climate change has emerged as a new overarching stress for the natural world, which in turn leads to the emergence of combinations of specific stresses and causes alteration in the severity of the other stress factors. Environmental changes leading to biological adaptation occur at different levels of biological organization, but understanding how the phenomenon of climate change affects the adaptive responses of ecosystems is still limited, especially in the light of interplay between chemical/pollutant stressors, land use changes, and changing climatic regimes (Niinemets et al., 2017).

7.6 LIPID PROFILE

Algae are characterized in part by the diversity of food reserves found in their cells (Barsanti et al., 2008). Among the algae, diatoms (Bacillariophyceae) are special because of their ornate siliceous frustules and their storage of lipids as their reserve food, which are clearly visualized as discrete lipid bodies (LBs) under light microscopy (Figure 7.2). Because diatoms are microscopic and have a short life span, even a small-scale change can affect their life cycles, which may be manifested as a change in its lipid characteristics; in particular, a change in the number and biovolume of LBs (which can be easily seen with microscopic investigation). Documentation of LB status in diatoms under different types of stresses is promising as an indicator of diatom stress and/or health. For example, Evans (1960) reported that diatom cells (*Pinnularia biceps* f. *minutissima*) at the edges of a pond that had large oil globules survived better than cells with fewer lipids. LBs status under changing climate conditions play a significant role as a proxy to understand the impact of climate change. Increased LBs may help cells to combat unfavorable conditions and aid in their survival. In contrast, cells lacking LBs may be more easily debilitated under poor conditions and unable to recover when provided with improving conditions. Under nitrogen starvation, induction of LBs is often apparent (Jiang et al., 2012) at the cost of a lower growth rate. Environmental disturbances are also associated with LB induction and modification of chloroplasts (Julius and Theriot, 2007). Published reports indicate that climate change, especially increased carbon dioxide concentration in the atmosphere, can be partly mitigated by fertilizing oceanic water with iron (Martin et al., 1989). Iron is an essential nutrient for the growth of phytoplankton and is a limiting nutrient in some oceanic waters, which results in restricted growth of phytoplankton and prevents bloom formation. Because marine phytoplankton are chiefly dominated by diatoms, research on the effect of iron on the diatoms is needed, but is understudied. Published reports on the effects of a variety of metals on diatoms in various aquatic environments indicate that diatoms are sensitive to metal toxicity for some metals, which may be manifested as alteration in their lipidomic profile. For example, Cu stress (in laboratory conditions) resulted in a higher lipid content in *Haslea ostrearia* in comparison to the control (Joux-Arab et al., 2000), indicating LB changes in response to metal

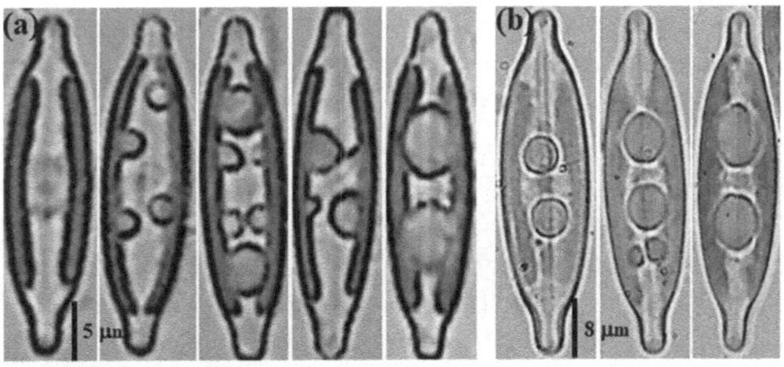

FIGURE 7.2 Globular lipid bodies inside (a) Navicula sp.1 and (b) Navicula sp. 2.

stress. In the same context, Pandey and Bergey (2016) found LB induction in terms of both the number and total biovolume of LBs for several different diatom species collected from metalliferous (Cu and Zn) polluted sites in Rajasthan, India. In contrast, Fe limitation in *Pseudo-nitzschia* had little effect in terms of lipid profile, whereas Cu-limited exposure produced increased lipid content in the same diatom species (Lelong et al., 2013).

Lipid molecular profiling of diatoms has emerged as a potential proxy for understanding paleoclimatic history of polar regions with special reference to climate change (Crosta et al., 2021; Vorrath et al., 2019, 2020). For example, analysis of sea-ice diatoms from the Antarctic marine sediments and the Southern Ocean showed the presence of di-unsaturated highly branched isoprenoid (HBI) and alkene (HBI diene, C25:2) (Massé et al., 2011). Similarly, Belt et al. (2016) reported the dominance of the sea ice diatom *Berkeleya adeliensis* in the platelet ice of the Antarctic region along with the presence of HBI alkene. Belt et al. (2016) found that super-cooled freshwater inflow is strongly associated with platelet ice formation. Therefore, they hypothesized that discharges from nearby glaciers and ice shelves influenced the level of sedimentary diene II in the Antarctic region, which showed the importance of organic biochemical lipid biomarkers in understanding past paleoclimatic conditions in the Antarctic region. Similarly, the sterol brassicasterol (24-methylcholesta-5, 22-dien-3β-ol) found in several diatom species (especially pennate diatoms) has also been used as a proxy for estimating diatom abundances (Calvo et al., 2004; Rampen et al., 2010). Calvo et al. (2011) examined CO_2-related changes in the eastern Equatorial Pacific since the last glacial period. In this study, they found that higher values of a marine lipidomic biomarker [i.e., the brassicasterol/(brassicasterol+alkenone) ratio] indicates the dominance of diatoms over coccolithophores. From the above discussion, it is clear that organic lipid biomarkers from diatom communities can be a sensitive and valuable tool for understanding the glaciated and deglaciated cover of polar regions, and hence in understanding the impact of climate change in polar regions.

7.7 SILICA PRODUCTION

Silica is the second most abundant element after oxygen in the earth's crust. Although abundant, the role of silica in biological systems is highly variable. Silica is normally released by the weathering of rocks in the soil, and plants take up this soil-associated silica. Silica is taken up by plants, and the silica content of plants varies among species. Silica is especially common in grasses, where it can protect against grazing by herbivores. With the decay of plants, silica enters fluvial ecosystems, and from there, the dissolved silica reaches estuaries and finally the ocean.

Diatoms are characterized by a high silica content, with silica making up approximately 70% of the dry weight of diatoms. This silica forms a protective cell wall known as the frustule. Diatoms form their siliceous frustules from dissolved silicic acid. Despite their relatively heavy frustules, lighter weight cell contents, including LBs, allow planktonic diatoms to remain in the water column (Villareal, 1988), but this heavy frustule also aids their export to the deep ocean after cell death. The unique morphological and physiological attributes of diatoms establish an important link between the marine carbon and silica cycles.

This terrestrial to aquatic biogenic silica cycle may be impacted by climate change. For example, reduced snow cover in northern woodlands results in more episodes of soil freezing, which freezes the shallow fine roots of sugar maples (*Acer saccharum*), significantly reducing silica uptake and, scaling up, can reduce the export of biogenic silica from the ecosystem (Maguire et al., 2017). Such reduced terrestrial exports of silica would lead to less availability of silica in the ocean and hence affect the growth and physiology of diatom communities – possibly leading to a decrease in overall diatom abundance (Goffart et al., 2002). Reductions in diatom abundance would impact oceanic food webs and could reduce CO_2 uptake in the oceans. For example, Javaheri et al. (2015) examined the silicification process in marine diatom

Thalassiosira pseudonana in batch culture under three different temperature regimes (14, 18 and 23 °C). Diatoms showed temperature-dependent growth. Additionally, they found that temperature affected the silicification patterns; that is, silicon-rich cultures formed mesh-like patterns, whereas silicon-limited cultures developed tree-like patterns with lower silicification under the influence of higher temperature. In contrast to potential silica dynamics in forests, Hawkings et al. (2018) describes higher dissolved silica concentrations in glacier meltwater from the Leverett Glacier, a large (~600 Km^2) catchment of the Greenland Ice Sheet versus non-glacier-fed streams. In their model, Hawkings et al. (2018) found that diatom productivity was significantly governed by changes in Si availability (through ice melt water) in ocean water during glaciation and deglaciation processes. Results were compared to those of a high-latitude lake sediment core from North America, which indicated relatively low Si release in comparison to the Greenland glacier meltwater (Hawkings et al., 2018). These findings strongly support the view that paleo-ice sheet growth and decay affect oceanic delivery of Si and that glacier melting in Greenland can stimulate the productivity of diatom communities and perhaps lower CO_2 levels in this region.

The process of silicification in diatoms under changing pH (oceanic acidification) has been little explored. Petrou et al. (2019) reported that silica incorporation by diatoms diminishes as waters become more acidic in the Southern Ocean, where diatoms are important components of both the phytoplankton and the community growing under sea ice. The projected CO_2 levels and the resultant ocean acidification will affect these ocean diatom communities by 2100, with differential effects among species (Petrou et al., 2019). From this study, it is clear that if the rate of oceanic acidification continues, the balance between carbon and silicon cycles will be altered, as will the composition of diatom assemblages, causing alteration in diatom cell ballasting and finally affecting the vertical flux of diatoms into the deep ocean. Kuefner et al. (2020) analyzed diatom communities and corresponding environmental parameters from 41 lakes encompassing a wide altitudinal range in the Bavarian and Tyrolican Alps. Statistical analysis showed that temperature and altitudinal position of the lakes were decisive factors for relating the "silicification value" of frustules (intensity of silicification in benthic diatoms), whereas the presence of macrophytes and trophic status of the lake had little impact. This finding showed that the silicification process in diatoms is correlated with temperature and therefore indicates that the silicification value might be useful as a proxy to assess the impact of climate change on diatoms in freshwater lakes.

Oceanic acidification may have a positive or negative effect on the ecology and physiology of diatom communities. On the basis of an Earth system model stimulation of a five-mesocosm experiment with natural plankton communities from different marine biomes, Taucher et al. (2022) found that oceanic diatom populations will decrease in the upcoming century due to an increase (17±6%) in Si:N export to the deep ocean (due to slower chemical dissolution of silica), resulting in a decrease in $Si(OH)_4$ concentration of surface oceanic waters. Oceanic acidification due to climate change is the main driving force for this alteration, triggering a projected global decline of diatoms by 13–26% by the year 2200.

7.8 PERSPECTIVES

Continuing climate change affects all spheres of life, but current and future effects on biological communities (especially primary producers) are incompletely understood. Because diatoms are the dominant flora of the oceans, the effects of climate change on diatoms are of considerable importance. From this chapter, it is clear that climate change affects diatoms at different ecological levels (i.e., individual, population, and community levels), which is mainly manifested as alteration in diatom abundance, bloom formation, size reduction, changes in life-forms, and frustule silica content. Along with this, molecular-level changes in the form of organic lipid biomarkers of diatoms may be useful indicators of climate change (Figure 7.3).

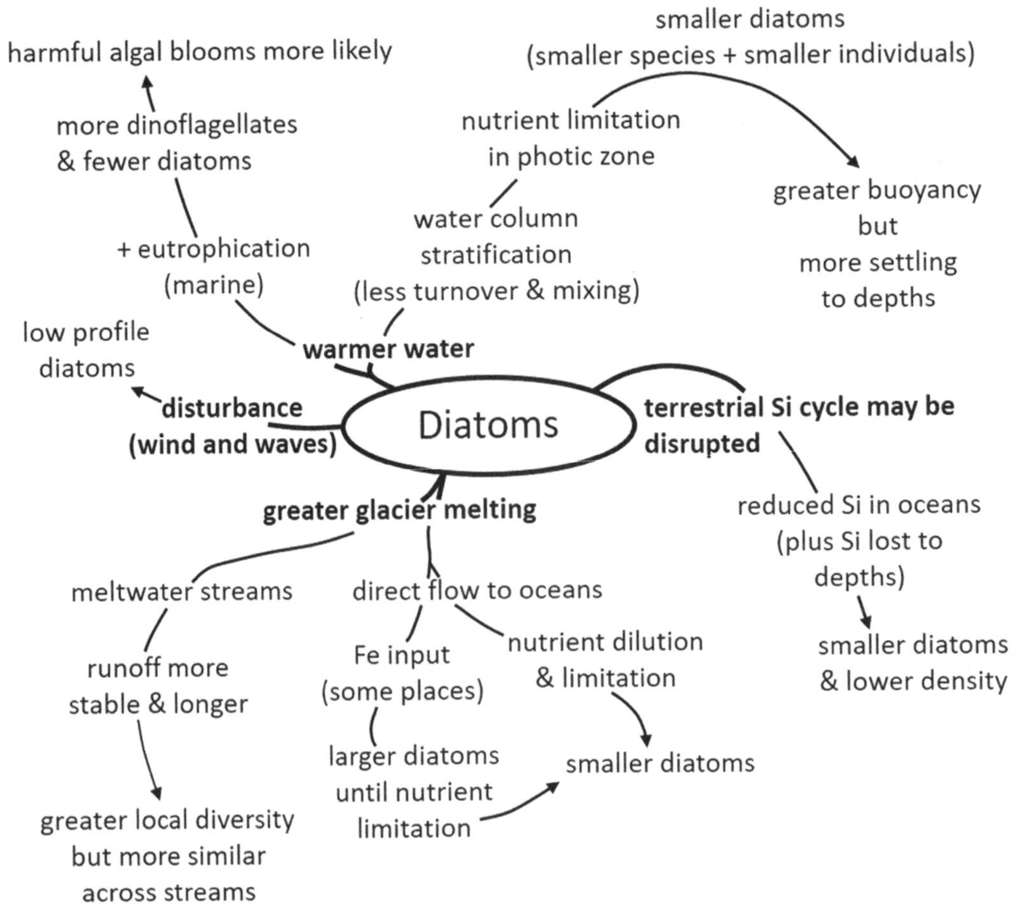

FIGURE 7.3 Flow chart showing the impact of climate change on diatoms.

7.9 CONCLUSIONS

Finally, we conclude that more comprehensive research on diatoms is urgently required in order to understand the impact of climate change on these primary producers, which will also be very helpful in mitigating the undesirable effects of climate change on biological organisms.

7.10. ACKNOWLEDGEMENTS

This work was supported by UGC-BSR Start-Up-Grant for newly recruited faculty at Assistant Professor level in Science Departments of various universities (2019–2020) (File No.30-494/2019 (BSR) and SERB-SRG (File No. SRG/2020/000432). Sangeeta thanks SERB, New Delhi for financial assistance in the form of JRF.

REFERENCES

Abonyi et al. (2020). Cell size decrease and altered size structure of phytoplankton constrain ecosystem functioning in the middle Danube River over multiple decades, *Ecosystems*, 23, 1254–1264. 10.1007/s10021-019-00467-6

Adams et al. (2013). Diatoms can be an important exception to temperature–size rules at species and community levels of organization, *Global Change Biology*, 19, 3540–3552. 10.1111/gcb.12285

André et al. (2010). Modelling climate-change-induced nonlinear thresholds in cephalopod population dynamics, *Global Change Biology*, 16, 2866–2875. 10.1111/j.1365-2486.2010.02223.x

Atkinson D (1995). Effects of temperature on the size of aquatic ectotherms: exceptions to the general rule, *Journal of Thermal Biology*. 20, 61–74. 10.1016/0306-4565(94)00028-H

Barral-Fraga et al. (2016). Short-term arsenic exposure reduces diatom cell size in biofilm communities, *Environmental Science and Pollution Research*, 23, 4257–4270.

Barsanti et al. (2008). The world of algae. In: Evangelista, V., Barsanti, L., Frassanito, A.M., Passarelli, V., Gualtieri, P. (Eds.), *Algal Toxins: Nature, Occurrence, Effect and Detection*. Springer Science-Business Media B.V, Netherlands, pp. 1–15.

Belt et al. (2016). Source identification and distribution reveals the potential of the geochemical Antarctic sea ice proxy IPSO25, *Nature Communication*, 7, 12655. 10.1038/ncomms12655

Bopp (2005). Response of diatoms distribution to global warming and potential implications: A global model study, *Geophysical Research Letters*, 32 (L18606). 10.1029/2005GL023653

Bramburger et al. (2017). Decrease in diatom cell size during the 20th century in the Laurentian Great Lakes: a response to warming waters? *Journal of Plankton Research*, 39, 199–210. 10.1093/plankt/fbx009

Brown et al. (2004). Toward a metabolic theory of ecology, *Ecology*, 85, 1771–1789. 10.1890/03-9000

Cattaneo et al. (2004). Diatom taxonomic and morphological changes as indicators of metal pollution and recovery in Lac Dufault (Quebec, Canada), *Journal of Paleolimnology*, 32, 163–175. 10.1023/B:JOPL.0000029430.78278.a5

Calvo (2004). Dust-induced changes in phytoplankton composition in the Tasman Sea during the last four glacial cycles, *Paleoceanography*, 19, PA2020. 10.1029/2003PA000992

Calvo et al. (2011). Eastern Equatorial Pacific productivity and related-CO2 changes since the last glacial period, *Proceeding of National Academy of Sciences*, 108(14), 5537–5541. 10.1073/pnas.1009761108

Cavicchioli et al. (2019). Scientists' warning to humanity: microorganisms and climate change, *Nature Reviews Microbiology*, 17, 569–586. 10.1038/s41579-019-0222-5

Costa et al. (2021). Large diatom bloom off the Antarctic Peninsula during cool conditions associated with the 2015/2016 El Niño, *Communications Earth and Environment*. 2, 252. 10.1038/s43247-021-00322-4

Crosta et al. (2021). Multi-decadal trends in Antarctic sea-ice extent driven by ENSO–SAM over the last 2000 years, *Nature Geoscience*, 14, 156–160. 10.1038/s41561-021-00697-1

Daufresne et al. (2009). Global warming benefits the small in aquatic ecosystems, *Proceedings of the National Academy of Sciences*, 106, 12788–12793. 10.1073/pnas.0902080106

Durant et al. (2007). Climate and the match or mismatch between predator requirements and resource availability, *Climate Research*, 33, 271. 10.3354/cr033271

Evans (1960). Further investigations of the algae of pond margins, *Hydrobiologia*, 15, 384–394. 10.1007/BF00046420

Fell et al. (2018). Declining glacier cover threatens the biodiversity of alpine river diatom assemblages, *Global Change Biology*, 24, 5828–5840. 10.1111/gcb.14454

Finkel et al. (2009a). Phytoplankton in a changing world: cell size and elemental stoichiometry, *Journal of Plankton Research*, 32,119–137. 10.1093/plankt/fbp098

Finkel et al. (2009b). Environmental control of diatom community size structure varies across aquatic ecosystems, *Proceedings of the Royal Society B: Biological Sciences*, 276, 1627–1634. 10.1098/rspb.2008.1610

Fleming et al. (2018). *Coastal effects*. In Reidmiller, D.R., C.W. Avery, D.R. Easterling, K.E (eds.), *Impacts, risks, and adaptation in the United States: Fourth national climate assessment, volume II*. https://nca2018.globalchange.gov/.

Forster et al. (2012). Warming-induced reductions in body size are greater in aquatic than terrestrial species, *Proceedings of the National Academy of Sciences*, 109, 19310–19314. 10.1073/pnas.1210460109

Gardner et al. (2011). Declining body size: a third universal response to warming? *Trends in ecology & evolution*, 26, 285–291. 10.1016/j.tree.2011.03.005

Gillooly et al. (2001). Effects of size and temperature on metabolic rate, *Science*, 293(5538), 2248–2251. 10.1126/science.1061967

Gobler, C.J. (2020). Climate Change and Harmful Algal Blooms: Insights and perspective, *Harmful Algae*, 91, 101731. 10.1016/j.hal.2019.101731

Goffart et al. (2002). Changes in the development of the winter-spring phytoplankton bloom in the Bay of Calvi (northwestern Mediterranean) over the last two decades: a response to changing climate. *Marine Ecology Progress Series*, 236, 45–60. 10.3354/MEPS236045

Hawkings et al. (2018). The silicon cycle impacted by past ice sheets, *Nature Communications*, 9, 3210. 10.1038/s41467-018-05689-1

Hooper et al. (2005). Effects of biodiversity on ecosystem functioning: a consensus of current knowledge, *Ecological Monographs*, 75(1), 3–35. 10.1890/04-0922

IPCC (2007). Intergovernmental Panel on Climate Change (IPCC). Climate change 2007: synthesis report. Geneva, Switzerland. http://www.ipcc.ch/publications_and_data/ar4/syr/en/main.html.

IPCC (2018). Summary for Policymakers. In Masson-Delmotte, V., P. Zhai, H.-O. Pörtner, D. Roberts, J. Skea, P.R. Shukla, A. Pirani, W. Moufouma-Okia, C. Péan, R. Pidcock, S. Connors, J.B.R. Matthews, Y. Chen, X. Zhou, M.I. Gomis, E. Lonnoy, T. Maycock, M. Tignor, and T. Waterfield (eds.), *Global Warming of 1.5°C. An IPCC Special Report on the impacts of global warming of 1.5°C above pre-industrial levels and related global greenhouse gas emission pathways, in the context of strengthening the global response to the threat of climate change, sustainable development, and efforts to eradicate poverty*. Cambridge University Press, Cambridge, UK and New York, NY, USA, pp. 3–24. 10.1017/9781009157940.001

Irwin and Oliver (2009). Are ocean deserts getting larger? *Geophysical Research Letter*, 36(18), L18609. 10.1029/2009gl039883

Javaheri et al. (2015). Temperature affects the silicate morphology in a diatom, *Scientific Reports*, 5, 11652. 10.1038/srep11652

Jiang et al. (2012). Photosynthetic performance, lipid production and biomass composition in response to nitrogen limitation in marine microalgae, *Plant Physiology and Biochemistry*, 54, 70–77. 10.1016/j.plaphy.2012.02.012

Jin and Agustí (2018). Fast adaptation of tropical diatoms to increased warming with trade-offs, *Scientific Reports*, 8, 17771. 10.1038/s41598-018-36091-y

Joux-Arab (2000). Do toxicity and accumulation of copper change during size reduction in the marine pennate diatom *Haslea ostrearia*? *Marine Biology*, 136, 323–330. 10.1007/s002270050690

Julius and Theriot (2007). The diatoms: a primer. In: Smol, J.P., and E.F. Stoermer (Eds.), *The Diatoms: Applications for the Environmental and Earth Sciences*, second eds. Cambridge University Press, London, pp. 8–22. 10.1017/CBO9780511763175.003

Kerrigan et al. (2015). Community- and population-level changes in diatom size structure in a subarctic lake over the last two centuries, *PeerJ*, 3, e1074. 10.7717/peerj.1074

Kilham et al. (1996). Linking planktonic diatoms and climate change in the large lakes of the Yellowstone ecosystem using resource theory, *Limnology and Oceanography*, 41, 1052–1062. 10.4319/lo.1996.41.5.1052

Kong et al. (2020). Unravelling winter diatom blooms in temperate lakes using high frequency data and ecological modelling, *Water research*, 190, 116681. 10.1016/j.watres.2020.116681

Kuefner et al. (2020). The silicification value: a novel diatom-based indicator to assess climate change in freshwater habitats, *Diatom Research*, 35(1), 1–16. 10.1080/0269249X.2020.1722246

Lampe et al. (2018). Different iron storage strategies among bloom-forming diatoms. *Proceedings of the National Academy of Sciences of the United States of America*, 115(52), E12275–E12284. 10.1073/pnas.1805243115

Lazarus et al. (2014). Cenozoic Planktonic Marine Diatom Diversity and Correlation to Climate Change, *PLoS ONE*, 9(1), e84857. 10.1371/journal.pone.0084857

Lelong (2013). Iron and copper limitations differently affect growth rates and photosynthetic and physiological parameters of the marine diatom *Pseudo-nitzschia delicatissima*, *Limnology and Oceanography*, 58, 613–623. 10.4319/lo.2013.58.2.0613

Lewis (1976). Surface/volume ratio: implications for phytoplankton morphology, *Science*, 192(4242), 885–887. 10.1126/science.192.4242.885

Limburg et al. (2020). Ocean Deoxygenation: A Primer, *One Earth*, 2(1), 24–29. 10.1016/j.oneear.2020.01.001

Liao et al. (2020). Lake diatom response to climate change and sedimentary events on the southeastern Tibetan Plateau during the last millennium, *Quaternary Science Reviews*, 241, 106409. 10.1016/j.quascirev.2020.106409

Luís et al. (2011). Environmental impact of mining activities in the Lousal area (Portugal): chemical and diatom characterization of metal-contaminated stream sediments and surface water of Corona stream, *Science of the Total Environment*, 409, 4312–4325. 10.1016/j.scitotenv.2011.06.052

Maguire et al. (2017). Winter climate change and fine root biogenic silica in sugar maple trees (*Acer saccharum*): implications for silica in the Anthropocene, *Journal of Geophysical Research: Biogeosciences*, 122, 708–715. 10.1002/2016JG003755

Marchetti et al. (2009). Ferritin is used for iron storage in bloom-forming marine pennate diatoms, *Nature*, 457, 467–470. 10.1038/nature07539.
Martin (1990). Glacial-interglacial CO_2 change: The Iron Hypothesis, *Paleoceanography and Paleoclimatology*, 5, 1–13. 10.1029/PA005i001p00001
Martin et al. (1989). Vertex: Phytoplankton/Iron studies in the Gulf of Alaska, *Deep-Sea Research Part A. Oceanographic Research Paper*, 36, 649-580. 10.1016/0198-0149(89)90144-1
Martin, J., and Fitzwater, S. (1988). Iron deficiency limits phytoplankton growth in the north-east Pacific subarctic, *Nature*, 331, 341–343. 10.1038/331341a0
Massé et al. (2011). Highly branched isoprenoids as proxies for variable sea ice conditions in the Southern Ocean, *Antarctic Science*, 23, 487–498. 10.1017/S0954102011000381
Miklasz and Denny (2010). Diatom sinking speeds: Improved predictions and insight from a modified Stokes' law, *Limnology and Oceanography*, 55(6), 2513–2525. 10.4319/lo.2010.55.6.25132513
Morán et al. (2010). Increasing importance of small phytoplankton in a warmer ocean, *Global Change Biology*, 16, 1137–1144. 10.1111/j.1365-2486.2009.01960.x
Morin and Coste (2006). Metal-induced shifts in the morphology of diatoms from the Riou Mort and Riou Viou streams (South west France). In: Acs, E., K.T. Kiss, J. Padiska, and K. Szabo (Eds.), *Use of Algae for Monitoring Rivers VI*. Hungarian Algological Society, God, Hungary, Balatonfured, pp. 91–106.
Morin et al. (2012). Consistency in diatom response to metal-contaminated environments. In: Guasch, H., A. Ginebreda, and A. Geiszinger (Eds.), *Emerging and Priority Pollutants in Rivers*. Springer-Verlag, Berlin, pp. 117–146.
Mousing et al. (2014). Global patterns in phytoplankton community size structure—evidence for a direct temperature effect, *Marine Ecological Progress Series*, 497, 25–38. 10.3354/meps10583
Niinemets et al. (2017). Interacting environmental and chemical stresses under global change in temperate aquatic ecosystems: stress responses, adaptation, and scaling, *Regional Environmental Change*, 17, 2061–2077. 10.1007/s10113-017-1196-3
Pandey and Bergey (2016). Exploring the status of motility, lipid bodies, deformities and size reduction in periphytic diatom community from chronically metal (Cu, Zn) polluted waterbodies as a biomonitoring tool. *Science of the Total Environment*, 550, 372–381. 10.1016/j.scitotenv.2015.11.151
Partridge and Coyne (1997). Bergmann's rule in ectotherms: Is it adaptive? *Evolution*, 51, 632–635. 10.2307/2411139
Peter and Sommer (2012). Phytoplankton cell size: intra- and interspecific effects of warming and grazing, *PloS one*, 7, e49632. 10.1371/journal.pone.0049632
Peters (1983). *The Ecological Implications of Body Size*. Cambridge University Press, New York. 10.1017/CBO9780511608551
Petrou et al. (2019). Acidification diminishes diatom silica production in the Southern Ocean, *Nature Climate Change*, 9, 781–786. 10.1038/s41558-019-0557-y
R Development Core Team (2010). *R: A Language and Environment for Statistical Computing*. R Foundation for Statistical Computing, Vienna, Austria.
Rampen et al. (2010). A comprehensive study of sterols in marine diatoms (Bacillariophyta): Implications for their use as tracers for diatom productivity, *Limnology and Oceanography*, 55, 91–105. 10.4319/lo.2010.55.1.0091
Reuman et al. (2013). A metabolic perspective on competition and body-size reductions with warming, *Journal of Animal Ecology*. 10.1111/1365-2656.12064
Riaux-Gobin and Poulin (2004). Possible symbiosis of Berkeleya adeliensis Medlin, *Synedropsis fragilis* (Manguin) Hasle et al. and *Nitzschia lecointei* Van Heurck (Bacillariophyta) associated with land-fast ice in Adélie Land, Antarctica, *Diatom Research*, 19, 265–274. 10.1080/0269249X.2004.9705874
Round et al. (1990). *The diatoms: Biology and morphology of the genera*. Cambridge University Press, Cambridge
Rüger and Sommer (2012). Warming does not always benefit the small – Results from a plankton experiment, *Aquatic Botany*, 97, 64–68. 10.1016/j.aquabot.2011.12.001
Schmittner (2005). Decline of the marine ecosystem caused by a reduction in the Atlantic overturning circulation, *Nature*, 434(7033), 628–633. 10.1038/nature03476
Sett et al. (2018). Shift towards larger diatoms in a natural phytoplankton assemblage under combined high-CO_2 and warming conditions, *Journal of Plankton Research*, 40(4), 391–406. 10.1093/plankt/fby018
Sienkiewicz et al. (2017). Unusual reaction of diatom assemblage on climate changes during the last millennium: a record from Spitsbergen lake, *Journal of Paleolimnology*, 58, 73–87. 10.1007/s10933-017-9962-2
Sheridan and Bickford (2011). Shrinking body size as an ecological response to climate change, *Nature Climate Change*, 1, 401–406. 10.1038/nclimate1259

Shi et al. (2010). Effect of ocean acidification on iron availability to marine phytoplankton, *Science*, 327, 676–679. 10.1126/science.1183517

Smetacek (1999). Diatoms and the ocean carbon cycle, *Protist*, 150, 25–32. 10.1016/S1434-4610(99)70006-4

Smol et al. (2005). Climate-driven regime shifts in the biological communities of arctic lakes. *Proceedings of National Academy of Sciences, USA*, 102(12), 4397–4402. 10.1073/pnas.0500245102

Smol and Stoermer (2010). *The Diatoms: Applications for the Environmental and Earth Sciences*. Cambridge University Press, New York.

Snoeijs et al. (2002). The importance of diatom cell size in community analysis, *Journal of Phycology*, 38, 265–272. 10.1046/j.1529-8817.2002.01105.x

Stockhecke et al. (2021). Temperature, precipitation, and vegetation changes in the Eastern Mediterranean over the last deglaciation and Dansgaard-Oeschger events, *Palaeography, Palaeoclimatology, Palaeoecology*, 577, 110535. 10.1016/j.palaeo.2021.110535.

Stelzer (2002). Phenotypic plasticity of body size at different temperatures in a planktonic rotifer: mechanisms and adaptive significance, *Functional Ecology*, 16, 835–841. 10.1046/j.1365-2435.2002.00693.x

Sunday et al. (2014). Evolution in an acidifying ocean, *Trends in Ecology & Evolution*, 29, 117–125. 10.1016/j.tree.2013.11.001

Svensson et al. (2014). Diatom Cell Size, Coloniality and Motility: Trade-Offs between Temperature, Salinity and Nutrient Supply with Climate Change, *PLoS ONE*, 9(10), e109993. 10.1371/journal.pone.0109993

Talling (2010). Potassium – a non-limiting nutrient in fresh waters? *Freshwater Reviews*, 3, 97–104. 10.1608/FRJ-3.2.1

Tarone et al. (2011). Population and Temperature Effects on *Lucilia sericata* (Diptera: Calliphoridae) Body Size and Minimum Development Time, *Journal of Medical Entomology*, 48, 1062–1068. 10.1603/me11004

Taucher et al. (2022). Enhanced silica export in a future ocean triggers global diatom decline, *Nature*, 605, 696–700. 10.1038/s41586-022-04687-0.

Thompson et al. (1992). Effects of variation in temperature I. On the biochemical composition of eight species of marine phytoplankton, *Journal of Phycology*, 28, 481–488. 10.1111/j.0022-3646.1992.00481.x

Villareal et al. (1988). Positive buoyancy in the oceanic diatom *Rhizosolenia debyana* H. Peragallo, *Deep-Sea Research*, 35, 1037–1045.

Virta and Teittinen (2022). Threshold effects of climate change on benthic diatom communities: Evaluating impacts of salinity and wind disturbance on functional traits and benthic biomass, *Science of the Total Environment*, 826, 154130. 10.1016/j.scitotenv.2022.154130.

Visser and Both (2005). Shifts in phenology due to global climate change: the need for a yardstick, *Proceedings of the Royal Society B*, 272, 2561–2569. 10.1098/rspb.2005.3356

Vorrath et al. (2019). Highly branched isoprenoids for Southern Ocean sea ice reconstructions: a pilot study from the Western Antarctic Peninsula, *Biogeosciences*, 16, 2961–2981. 10.5194/bg-16-2961-2019.

Vorrath et al. (2020). Sea ice dynamics in the Bransfield Strait, Antarctic Peninsula, during the past 240 years: a multi-proxy intercomparison study, *Climate of the Past*, 16, 2459–2483. 10.5194/cp-16-2459-2020

Vose et al. (2017). Temperature changes in the United States. In: *Climate science special report: Fourth national climate assessment, volume I* [Wuebbles, D.J., D.W. Fahey, K.A. Hibbard, D.J.].

White et al. (1999). The effect of temperature on experimental and natural chemical weathering rates of granitoid rocks, *Geochimica et Cosmochimica Acta*, 63, 3277–3291. 10.1016/S0016-7037(99)00250-1

Whitlock (2005). Combining probability from independent tests: the weighted Z-method is superior to Fisher's approach, *Journal of evolutionary biology*, 18, 1368–1373. 10.1111/j.1420-9101.2005.00917.x

Wilhelm and Adrian (2008). Impact of summer warming on the thermal characteristics of a polymictic lake and consequences for oxygen, nutrients and phytoplankton, *Freshwater Biology*, 53(2), 226–237.

Winder et al. (2009). Lake warming favours small-sized planktonic diatom species, *Proceedings of the Royal Society B: Biological Sciences*, 276, 427–435. 10.1098/rspb.2008.1200

Winder and Sommer (2012). Phytoplankton response to a changing climate, *Hydrobiologia*, 698, 5–16. 10.1007/s10750-012-1149-2

Woodward et al. (2010). Ecological Networks in a Changing Climate, *Advances in Ecological Research*, 42, 71–138. 10.1016/B978-0-12-381363-3.00002-2

Xiao et al. (2018). Warming and eutrophication combine to restructure diatoms and dinoflagellates, *Water Research*, 128, 206–216. 10.1016/j.watres.2017.10.051

Yom-Tov et al. (2010). Temperature trends and recent decline in body size of the stone marten Martes foina in Denmark, *Mammalian Biology - Zeitschrift für Säugetierkunde*, 75, 146–150. 10.1016/j.mambio.2008.10.005

Yvon-Durocher et al. (2010). The temperature dependence of the carbon cycle in aquatic ecosystems, *Advances in Ecological Research*, 43, 267–313. 10.1016/B978-0-12-385005-8.00007-1

Yvon-Durocher et al. (2011). Warming alters the size spectrum and shifts the distribution of biomass in freshwater ecosystems, *Global Change Biology*, 17, 1681–1694. 10.1111/j.1365-2486.2010.02321.x

Zhang et al. (2017). Effect of the Silica Content of Diatoms on Protozoan Grazing, *Frontiers in Marine Sciences*, 4, 202. 10.3389/fmars.2017.00202

8 Diatoms

Systematics to Nanobiotechnology

Rahul Bose and Ruma Pal
Phycology and Algal Biotechnology Laboratory, Department of Botany, University of Calcutta, 35, Ballygunge Circular Road, Kolkata, West Bengal, India

CONTENTS

8.1 INTRODUCTION

Diatoms are unicellular eukaryotic photosynthetic algae, classified under the division Bacillariophyta (Round et al., 1990). Diatoms are believed to have developed as a result of a secondary endosymbiotic association between a red eukaryotic alga and a heterotrophic flagellate around 200 million years ago (Medlin et al., 2000). The earliest alga identified, according to a report published in 1703 by the Royal Botanical Society of London, was an epiphytic diatom (*Tabellaria flocullosa*) attached to the roots of the aquatic weed *Lemna* (Round, 1981). Diatoms are widespread in occurrence, ranging from freshwater to marine environments in planktonic or benthic forms, with a very high rate of productivity. Following the annual death of diatom blooms, a significant number of frustules are deposited at the ocean's bottom and combine to form diatom oozes, which ultimately fossilize and produce Diatomaceous earth (Hoek et al., 1995). Diatomaceous earth formation is governed by high diatom yields supported by adequate availability of light and dissolved nutrients, especially soluble silica (Harwood, 1999). Based on evidence from the fossil records, diatoms are thought to have evolved after the Permian-Triassic mass extinction basically in the Triassic period (250 million years ago) though earliest fossil recorded dating back about 180 million years ago (Armbrust, 2009; Kooistra et al., 2007; Sims et al., 2006).

Diatoms are dominated among phytoplankton communities in well-mixed upwelling and coastal regions as well as in and along the edge of sea ice. Their growth is controlled by several factors viz. availability of sufficient light, inorganic nitrogen, phosphorus, silicon, and trace elements (Morel & Price, 2003). They flourish mainly in silicate-rich environments, thereby depleting silicate concentration after bloom formation. Therefore, silica availability has received a lot of attention for diatom abundance (Lund, 1965; Lund & Reynolds, 1982; Paasche, 1980; Reynolds, 1984). In polar

DOI: 10.1201/9781003322115-8

environments, diatoms are critical components of the food webs supporting marine and terrestrial ecosystems because glaciers and permafrost limit photosynthesis on land. Diatoms dwell throughout the water column in the context of accessibility of light and nutrients for photosynthesis. Likewise, in high light and the nutrient-depleted zone they photosynthesize mainly at the surface of the ocean, whereas in nitrate-rich medium, they carry out the process even at a depth of around 100 m (Villareal et al., 1999).

Diatoms account for more than 20% net primary productivity of the marine ecosystem and 40% of the ocean carbon fixation (Mann, 1999; Nelson et al., 1995). Lipid is the major carbon storage molecule in diatoms, accounting for 15–25% of dry biomass through triacylglycerides (TAGs) and fatty acids (Mangas-Sánchez & Adlercreutz, 2015). Fucoxanthin is the most abundant carotenoid in diatoms, where it coexists with chlorophyll a, c, and chlorophyll a/c binding protein; therefore, it gives the diatom cell its golden-brown color (Gammone & D'Orazio, 2015; Zhang et al., 2015). Fucoxanthin is reported to have greater antioxidant properties due to the presence of an allenic link (-C=C-) in its structure (Sachindra et al., 2007).

In many aquatic ecosystems diatoms play a role of natural feedstock, containing important pigments, fatty acids, vitamins, and polysaccharides, among other substances. Polyunsaturated fatty acids (PUFAs) account for about 25% of diatoms' total fatty acid content, and they are considered to be a rich source of fish oil in aquatic biota. As a consequence, the use of selected strains of diatoms in aquaculture farms, like *Chaetoceros gracilis*, *Skeletonema costatum*, *Chaetoceros muelleri*, *Thalassiosira pseudonana,* and other diatoms, has been in practice for many years (Brown, 1991; Brown & Miller, 1992; Brown et al., 1999; Dunstan et al., 1993; Lombardi & Wangersky, 1995; Tam et al. 2021).

Due to the presence of nano-silica in diatom frustules, these unicellular algae work as tiny silica nanofabrication factories with multifunctional capabilities (Losic et al., 2009). Nanotechnologists and material scientists are interested in micro- to nanoscale applications of diatom frustules in nano-biotechnological domains. Through chemical and mechanical processing, nanostructured silica can be produced under controlled conditions for a variety of uses. The attachment of the desired biomolecule (enzymes, DNA, or antibodies) to this nanopore structure offers enormous promise for the development of label-free analytical tools or the improvement of catalytic characteristics. Additionally, it has been demonstrated that silica can be used to combine enzymes and DNA (oligonucleotides) (Losic et al., 2005; Zamora et al., 2009). Compared to conventional immobilization techniques, the encapsulating of enzymes in diatom biosilica demonstrates better enzymatic characteristics (Kato et al., 2020). The most abundant source of diatom bio-silica is diatomaceous earth (DE) with its various applications. Together with many filtering processes, in water filtration systems diatomaceous earth is used with varied permeability, governed by the different pore sizes of the frustules (Ongerth & Hutton, 1997; Schuler et al., 1991). Moreover, it may be transformed to dimagnesium silicide (Mg_2Si) in order to produce extremely promising thermoelectric nano-materials. Several other applications of diatom frustules involve light harvesting, device manufacture, photonics, molecule separation, sensing, and drug delivery systems, among many more. Because of their photoluminescence feature, diatoms are used to construct hundreds of biosensors on a microchip-sized platform, depending on the frustules structure and surrounding environment (Butcher et al., 2003; Bradbury, 2004; Dempsey et al., 1997; Losic et al., 2009; Meyer et al., 1995; Szczech & Jin,2008). Additionally, the formation of diatom biosilica and metal nano-composites as a result of metal exposure is also beneficial for a variety of biotechnological applications (Roychoudhury et al., 2016; Roychoudhury et al., 2021; Seckbach & Kociolek, 2011).

8.2 DIATOM TAXONOMY

Diatom taxonomy has faced tremendous challenges from time to time due to very minute sizes of the frustules with very particular genetically controlled ultrastuctural patterns that are considered for their identification. Early records on the structure, life cycle, and taxonomy of diatoms include the works of Geitler (1932), Hustedt (1927–1966), Kützing (1834a, 1834b, 1844), Karsten (1928), Pfitzer (1871), Schmidt (1874–1959) and Schutt (1896). Krammer & Lange-Bertalot's Süsswasserflora von Mitteleuropa (1986, 1988, 1991a, 1991b, 2000, 2004) was frequently used for identifying freshwater diatoms. Hasle & Syvertsen (1997) described marine planktonic diatoms solely from temperate and polar regions in their handbook. The majority of books in the series, such as ***Bibliotheca Diatomologica***, ***Iconographia Diatomologica***, ***Diatom Monographs*** and ***Diatoms of Europe,*** dealt with the taxonomy and biodiversity of the diatom flora (Levkov, 2009; Metzeltin & Lange-Bertalot, 2007).

Taxonomists who study diatoms are in a good position since the abundance of detail visible in the silica frustules offers a variety of features and states as a basis for defining taxa and determining their relationships to one another. Contrary to characteristics of sexual reproduction, these features are always accessible for research, and they may be fairly quickly investigated with electron microscopy (Round et al., 1990) or even only light microscopy (Hasle & Syvertsen, 1997). Based on how the pattern of ribs and pores is organized on the valve face, taxonomists have divided diatoms into broad groupings. The ribs of centric diatoms typically radiate out from a ring (annulus). The annulus is essentially circular and central in the majority of what are often referred to as "radial centrics," and the valve shape is circular. The annulus, however, may or may not be central and it need not be circular in "bipolar" or "multipolar" centrics (Crawford et al., 1994; Pickett-Heaps et al., 1990). The pennate diatoms are another important type. They are arranged like a feather (Latin: "pinna" or "penna") and have an extended rib- or bar-like pattern center called the sternum from which the ribs stretch out somewhat perpendicularly. Pennate diatoms typically have elongated shapes, and their valves imitate ships. Araphid pennates lack these slits but frequently have labiate processes at the apices or centres. Raphid pennates feature one or two raphe slits within or closely linked with their sternum.

For the first time, diatoms were categorized by Karsten (1928) as belonging to the division (or phylum) Bacillariophyta. He distinguished two orders based on the structure of the pattern on the valves: the Centrales (central diatoms) and the Pennales (pennate diatoms). This split was made on the basis of the two orders' sexual reproductive features; centrales are oogamous, whereas pennales show isogamous with non-motile gametes. Silva (1962) later renamed the Karsten's bipartite classification Centrales and Pennales to Centrobacillariophyceae and Pennatibacillariophyceae, respectively. Following that, Round et al. (1990) classified the division Bacillariophyta into three primary groups based on valve symmetry and the presence of raphe systems: Coscinodiscophyceae, Fragilariophyceae, and Bacillariophyceae. The Coscinodiscophyceae family had individuals with a radial layout of the main valve pattern, similar to Centrales (Centrobacillariophyceae). The Fragilariophyceae and Bacillariophyceae were closely related to the Pennales (Pennatibacillariophyceae) of earlier categorization systems; both possessed transverse ribs and rows of holes covered by a longitudinal sternum. The two groups were defined by the lack (Fragilariophyceae) or presence (Bacillariophyceae) of a raphe system. However, Round et al. (1990) avoided accusing the three groups of polyphyly. Two of the three groups (Coscinodiscophyceae and Fragilariophyceae) were most likely paraphyletic, indicating that they did not adequately reflect key elements of diatom evolution. Medlin & Kaczmarska (2004) revised the tripartite classification of Round et al. (1990) based on molecular data of SSU rRNA sequences and proposed two subdivisions, namely Coscinodiscophytina and Bacillariophytina, followed by three classes viz. Coscinodiscophyceae, Mediophyceae, and Bacillariophyceae (CMB hypothesis). Among the proposed classifications, all diatoms with radially symmetric shapes were assigned to the Coscinodiscophyceae class, while only bipolar centric shapes and the Order Thalassiosirales were assigned to the Mediophyceae group, and finally all pennate forms of diatoms were assigned to the Bacillariophyceae class. However, a parallel study was conducted in order to classify diatoms at a higher level as well. Cavalier-Smith & Chao (1996) revised the phylogenetic classification of phylum Ochrophyta based on 18S rRNA sequences. Additionally, they have considered diatoms under sub-phylum Diatomeae and together with sub-phylum Phaeista constituted the

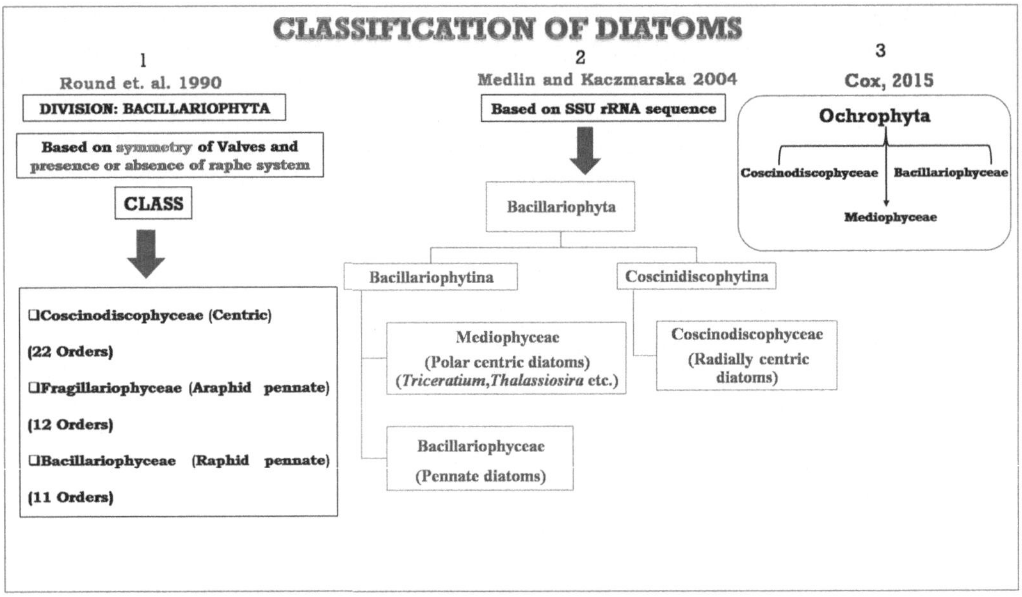

FIGURE 8.1 Classification of diatoms.

Ochrophytes group under kingdom Chromista. Later on, an amalgamated classification system was proposed by Cox (2015) by integrating the higher classification of Cavalier-Smith & Chao (1996) (Phylum: Ochrophyta) with the lower order of systematics like that of Medlin & Kaczmarska (2004). The world's biggest and most updated databases, Algaebase (https://www.algaebase.org/) and DiatomBase (https://www.diatombase.org/), both use Cox's suggested classification. The schematic representation of the same is illustrated in Figure 8.1.

8.3 DIATOMS IN NANOBIOTECHNOLOGY

Diatoms have been praised as a paradigm for bio-based nanostructured material because they build a hierarchical structure with nano-silica particles. They self-assemble into periodic micro- and nanoscale dimensions through a bottom-up, self-assembly process from soluble silicon under ambient conditions in a pH-dependent manner and being governed at a molecular level (Lechner & Becker, 2015; Pamirsky & Golokhvast, 2013; Sumper & Kroger, 2004; Wetherbee et al., 2000). Biomaterials have emerged as a fresh source of inspiration for the design and manufacture of superior nanostructured materials in the modern era. Recent breakthroughs in synthetic nanotechnology have been spurred by an almost continual global need for ever-smaller devices in the electrical, optical, chemical, and biological domains (Gordon et al., 2005) (Figure 8.2). Thus, the intricate structural geometric pattern of diatom frustules serves as a blueprint for modelling biomolecules at the micro- and nanoscale levels (Sumper & Brunner, 2006).

8.3.1 Diatoms in Lithography Studies

The silica frustules of diatoms are manipulated and transformed using metals and polymers for nano- and microfabrication for the production of inexpensive and environmentally benign lithographic templates (Losic et al., 2007). Material engineers have used various diatom shapes for etching lithography with nano-textured polymers. Use of diatoms as a base-medium for lithography requires two distinct methods.

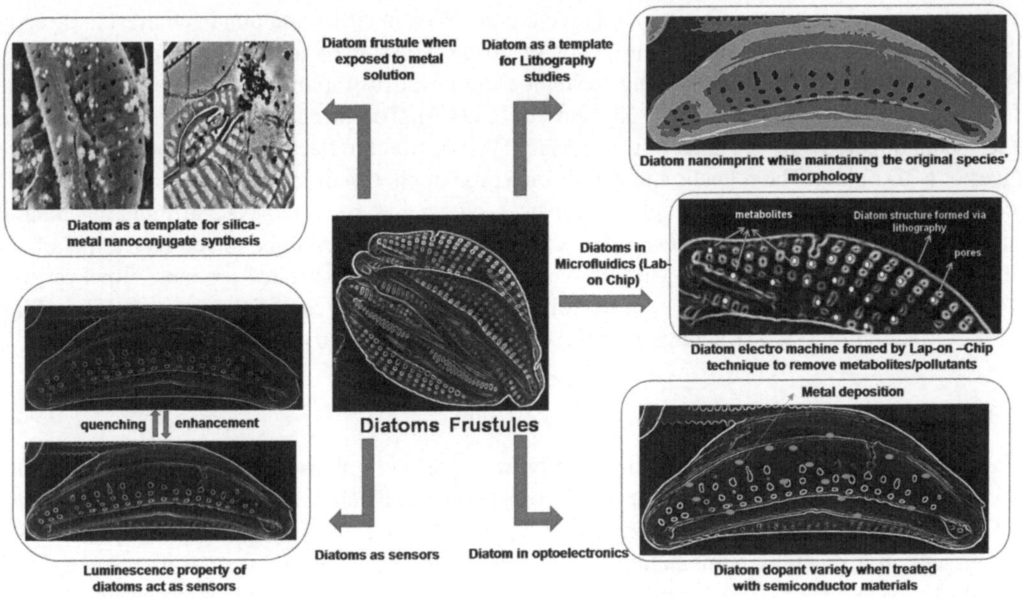

FIGURE 8.2 Application of diatom frustules.

The first one, **Nanoimprint lithography (NIL)**, employs polydimethylsiloxane (PDMS Sylgard 184) to imitate micro- and nanoscale patterns, which are immobilized on a glass surface covered with polymer in order to create negative duplicates of diatoms. Use of PDMS and the subsequent transfer of the soft PDMS pattern onto a hard UV-curable polymer such as NOA 60 (Norland Optical Adhesive 60) is used for replica production (Losic et al., 2007). These resultant replicas might be utilized for nanoscale patterning in a variety of devices, including biosensing devices, nanobioreactors, etc. Similarly, the micro- and nano-ordered arrays in species like *Cocconeis scutellum* var. *sullivanensis*, *Cocconeis scutellum* var. *posidoniae*, *Cocconeis pseudomarginata* may serve as templates for photolithographic techniques or serve as a model base for micro-channels (De Stefano et al., 2011). Carbothermal reduction was used to convert the biosilica templates into novel oxides or oxide-bearing composites while maintaining the original species' morphology (Breslin et al., 1995; Sandhage et al., 2002).

Positive and negative replicas of diatom-inspired structures are fabricated by means of **3D laser lithography (3DLL)**; to do so CAD (computer-aided design) software was basically used to create artificial diatom structures (Rhinoceros 4.0, McNeel) (Belegratis et al., 2013). By examining SEM pictures, the original designs were produced, and the matching CAD models were created. The 3D designs were encrypted onto a photosensitive material with great spatial precision through photon-induced polymerization. All artificial templates created were consistent, homogeneous, stable, and repeatable in terms of shape. The technique of constructing an artificial diatom template using 3DLL was restricted to micrometre to submicrometer scale dimensions. To overcome this problem, **Laser interference lithography (LIL)** is an alternative technique for patterning huge surfaces areas. LIL simultaneously scans regions from the submicron to cm^2 range slowly using numerous interfering laser beams. It also permits the production of a complex structure by organizing and imitating the architectures of diatoms with varying pore diameters as in the case of *Coscinodiscus wailesii* (Lasagni & Menéndez-Ormaza, 2010). Among the possible uses of bioinspired nanostructures derived from diatoms are mechanical protection (Hamm et al., 2003) and gas and nutrient exchange (Parkinson & Gordon, 1999).

In contrast to all photolithographic procedures, which involve chemical and heat processing, the utilization of protein hydrogels for biomimetic construction of 3D silica structures does not alter the volume of the templates. This method precipitates silica in vitro condition in contrast to directly employing biosilica from diatoms (Khripin et al., 2011). **Mask-directed multiphoton lithography (MDML)** technology is a 3D writing technique employed to rapidly make prototypes of user-defined designs (Kaehr et al., 2007; Nielson et al., 2009). The mechanism follows with a pulsed laser being directed onto a digital mirror device (DMD), which reflects a succession of 2D picture slices of a 3D object. The reflection then falls on a protein photosensitizer solution to generate a 3D protein hydrogel structure. Although significant progress has been achieved in imitating natural and manufactured diatom templates, relatively little work has been done to reproduce the silica-like characteristics of diatoms in-vitro employing native silica-associated proteins. However, the works of Khripin et al. (2011) have created diatom templates using the MDML approach, resulting in the formation of silica structures in artificial "diatoms" with a triangular shape.

8.3.2 Diatoms in Microfluidics (Lab-On-Chip)

The hierarchical structure and most importantly free hydroxyl groups on the surface of the diatom frustules are used in the development of microscale total analysis system (TAS) or lab-on-a-chip (LOC) platform. This kind of technology was shown to be successful at detecting many antibodies or antigens in a complex combination, and so has biological applications as well (Seckbach & Kociolek, 2011). Due to enrichment with negatively charged hydroxyl and amino groups, diatom frustules are sometimes treated with alkoxysilane-3 aminopropyltrimethoxy silane (APS) for immobilizing antibodies on the silica surface. Numerous studies have shown that diatom biosilica frustules functionalized with antibodies serving as a microscale biosensor platform for the selective, label-free detection of immunocomplex formation (Bismuto et al., 2008; De Stefano et al., 2009; Gale et al., 2009). Sometimes APS is labelled with the cross-linker N-5-azido 3-nitrobenzoyl oxysuccinimide (ANBNOS) due to absence of reactive functional groups. The cross-linker enables the antibodies to form strong bonds with the silica surface. Therefore, when IgY antibodies are treated with modified diatoms, they effectively bind (Townley et al., 2008). By oxidizing hydroxyl groups in the Fc region (fragment crystallizable area) of the antibodies to produce aldehydes, Townley et al. (2008) revealed that the carbohydrate moiety of antibodies may also be attached to the surface of diatoms. This technique also permits the binding of mouse anti-tubulin and goat anti-mouse IgG–HRP (immunoglobulin G horseradish peroxidase). In addition to binding single biomolecules, double antibodies may be labelled on the surface of the diatom and identified using FITC (fluorescein isothiocyanate) and confocal imaging. In reality, the modified silica walls of diatoms function as a μTAS (microscale total analysis system) containing antibody arrays for use in immuno-precipitation procedures.

8.3.3 Diatoms as Optical Sensors

Even though amorphous silica is the primary component of diatom frustules, varying amounts of other components such as clays, salts, organic matter, and inorganic oxides such as Al_2O_3, FeO_3, $CaCO_3$, CaO, K_2O, Na_2O, and MgO are also present. These contaminants have a significant impact on the light signal produced by diatoms. In order to eliminate these contaminants from frustules, many purification methods based on hot acid treatments were developed (Osman et al., 2009). The similarity between the porous silicon dioxide and the porous hydrated amorphous silica of diatom frustules promoted the latter to be considered as a natural gas transducer. In fact, there is a substantial contact between the vapour phase and the surface of the pore walls when a volatile material is entrapped in pores of nanometric dimensions. The gas may undergo a phase transition and become liquid through the tiny pores, whereas the silica matrix transforms into a mixed liquid–solid composite with significantly altered chemical and physical characteristics (Moretti et al., 2007a,b).

Due to the presence of reactive Si-OH (Silanol) groups on the valve surface of diatoms, the valve surface has sometimes been chemically changed using alkoxysilane compounds (R_3SiO-R) to form a Si–O–Si (Siloxane) bridge where the hydroxyl group of silanol condenses with the alkoxysilanes (Stewart & Buriak, 2000). The pore array of diatom frustules may function as a photonic crystal to promote wavelength-specific photon absorption. In addition, the process of ambient silica biomineralization introduces defects to the structure of biogenic silicon oxide, such as non-bridging oxygen centers (SiO:) and silanol (SiOH) groups. These imperfections reduce the conduction band of the material to permit photoluminescence, the spontaneous emission of blue light (450–500 nm) when excited by ultraviolet light with a greater energy. The hierarchical porosity and high specific surface area of newly synthesized diatom biosilica are favorable for sensor functions because they allow for the adsorption of large amounts of analyte macromolecules in a short period of time. It had been established by De Stefano and colleagues that the intrinsic photoluminescence (PL) of diatom biosilica in the wavelength range 450–690 nm was affected by the adsorption of organic molecules (acetone, ethanol, xylene, and pyridine) or inorganic molecules (NO_2) due to the electrophilic nature of the target molecules, which accepted electrons from the diatoms' silica backbone (Bismuto et al., 2008; De Stefano et al., 2005, 2009).

The surface silanol groups on the diatom biosilica pore array after functionalization with biomolecules, such as antibodies, serve as recognition elements for biosensing (Gale et al., 2009). Recent investigation demonstrated that silver nanoparticle-decorated diatom frustules increased the sensitivity of surface-enhanced Raman scattering (SERS) for detecting immunocomplex formation (Yang et al., 2015). SERS is a surface-sensitive method that amplifies Raman scattering by molecules adsorbed on metal nanoparticles through plasmon resonance. Initially, frustule biosilica was obtained from cultivated cells of the pennate diatom *Pinnularia* sp. The biogenic silica was then functionalized with amine groups that act as nucleation sites for the self-assembly of silver nanoparticles on the surface of the frustules. The antibody (goat anti-mouse IgG) was then linked to the silver nanoparticles through thiol groups, ionic, hydrophilic, and hydrophobic interactions. The antibody-functionalized diatom frustule was further treated with gold nanoparticle-capped mouse IgG molecules. The detection limit of mouse IgG molecules on diatom frustules was 100 times greater than that of the identical immunocomplex when immobilized onto a nonporous glass surface. In this case, the periodic sub-micron size pore structures of the diatom frustule were found to be responsible for the enhancement, which allowed guided-mode resonance at visible wavelengths to increase the SERS signals, in a manner similar to manufactured photonic crystals.

8.3.4 Diatoms as Precursor Material of Nanoconjugate Synthesis

The physicochemical synthesis of nanoparticles (NP) for commercial applications is often energy and time demanding, requiring high temperatures and pressures as well as hazardous/toxic chemical components that are all destructive to the environment. Temperature and hydrolysis time had a substantial effect on the average particle size, distribution, and form. As a consequence, there is an increasing need for technologies that enable the manufacture of nanostructures in a safer, more cost-effective, quicker, and more environmentally friendly manner (Bondareva et al., 2020; Farjadian et al., 2019; Kiani et al., 2020; Kiani et al., 2021; Tavakolizadeh et al., 2021). To produce these NPs, biological, renewable, and sustainable resources may be utilized in cost-effective and ecofriendly ways, obviating the need for arduous and costly processes (Ghotekar, 2019; Iravani, 2011; Iravani & Jamalipour, 2019; Iravani & Varma, 2019; Varma, 2012; Varma, 2019). Due to the frustule structure and rising abundance, diatoms in particular may be exploited as natural resources for eco-friendly synthesis of various nanostructures through metabolic insertion and in situ coating/fabrication (Gutu et al., 2009; Jeffryes et al., 2008; Qin et al., 2008; Schröfel et al., 2011; Yamazaki et al., 2010). Nevertheless, substantial technical challenges remain, including controlling the size, dispersity, stability, and shape of synthesized NPs, as well as their extraction and isolation with high purity, effectiveness, and functionality.

Various diatoms like *Diadesmis gallica*, *Stephanopyxis turris*, and *Navicula atomus* were cultivated in the bioreactor for the production of gold nanoparticles through extracellular polysaccharides released by diatom cells. Interestingly, these nanoparticles were discovered within diatom cells using Surface Enhance Raman Spectroscopy (SERS). Subsequently, it was also revealed that other than main pigments, various physiologically significant components such as hydroxyurea and retinol were responsible for nanoparticle production (Pytlik et al., 2017; Schröfel et al., 2011). Surprisingly, peptide-generated diatom frustules also facilitated the production of silver nanoparticles (NPs). The photoreduction-based NP synthesis kinetics was particularly sensitive and was reliant on peptide–diatom interactions that altered the rate of the synthesis (Gupta et al., 2018). Several species, including *Thalassiosira*, *Chaetoceros*, and *Skeletonema*, were employed to biofabricate silver nanoparticles with significant antibacterial activity against pathogens such as *Streptococcus pneumoniae*, *E. coli*, *Aeromonas sp.*, *Staphylococcus aureus*, and *Bacillus subtilis* (Mishra et al., 2020). Additionally, *Amphora copulata* biosynthesized spherical, hexagonal, and triangular gold-silica nanocomposites and the resultant nanocojugates were assessed for DNA-binding affinity using agarose gel electrophoresis. The particles were effectively attached to the DNA molecules without modifying their surfaces, which might have applications in biomedical research (Roychoudhury et al., 2016). Another genus of the same order, namely *Halamphora subturgida,* when treated with silver nitrate solution after acid hydrolysis resulted in the formation of nanoconjugate dendrites in a cost-effective way. The resultant nanoconjugate dendrites possessed fluorescence due to the intrinsic autoluminescence feature of the diatom frustules (Bose et al., 2021).

8.3.5 Diatoms in Optoelectronics

Making electroluminescent devices out of diatom biosilica after metabolically doped with germanium (Ge) is a feasible method of fabrication. The Ge oxide flaws in the biosilica act as phosphors, emitting light at specified wavelengths when an alternating current (AC) electric field is applied, and the pore structure aids in the culmination of the emission (Jeffryes et al., 2008). A wide variety of optoelectronic device applications, such as LED lighting and phosphor-based flat panel displays, rely on electroluminescence. Besides their applications in detecting nitric oxide (NO) (Bao et al., 2007), dihydrogen (H_2) (Fang et al., 2009), they were used as a template for biofabrication of electroluminescent material (Jeffryes et al., 2008), and fabrication of nanoelectronic devices (Bose et al., 2020).

The silica deposition procedure may be exploited to create a silica composite material, which in turn modifies the diatom frustule. Direct in-vivo integration of germanium into the frustules of *Nitzschia frustulum* resulted in a nanostructured Si-Ge oxide composite material with excellent mechanical properties (Qin et al., 2008, Rorrer et al., 2005). Similarly, incorporation of titanium (Ti) into the frustule biosilica was accomplished when dissolved Si in the form of Si $(OH)_4$ was fed to silicon-starved diatom cells in the presence of dissolved titanium in the form of Ti $(OH)_4$. The TEM and STEM-EDS techniques were used to examine whole biosilica frustules that contained metabolically implanted titanium. Titanium was selectively accumulated as a nanophase in the tiny pore structure along the base of each frustule pore. The diatom *Thalassiosira weissflogii* was shown to be capable of directly absorbing titanium (IV) bis-ammonium-lactato-dihydroxide (Ti-BALDH), which has a high solubility in water, and incorporating it into the frustule at a Si:Ti ratio of 0.34:1 in the girdle band, according to research (Lang et al., 2013). The findings also revealed that the diatom cells could grow and divide entirely on the Ti substrate, which was consistent with previous findings. Because of this research, the fascinating potential of using Ti substrates with high solubility for biomineralization in diatom cells is an active area of research.

The notion of employing diatom cell culture to take up soluble metals and concurrent integration of metals into the frustule biosilica was applicable to other metals as well, like aluminum (Machill et al., 2013), nickel (Townley et al., 2007, 2008), and europium (Zhang et al., 2015). In particular, the absorption of soluble europium (Eu) by the pennate diatom *Navicula* in cell culture was of special relevance for optoelectronic device applications (Zhang et al., 2015). A silicon-depleted diatom cell culture was treated with a combination of soluble silicon ($Na_2SiO_3.9H_2O$) and europium ($Eu(NO_3)_3.6H_2O$) in the ratio of 4:1 and incubated for 96 hours. Alcohol-soluble organic material from the diatom cells was removed, and the residual solid was thermally annealed in air at 1000°C for 30 minutes, resulting in the formation of Eu_2SiO_5 by solid-phase reaction.

There are not many inventions based on diatoms that have been documented or patented. Dickerson et al. (2007) patented manufacturing techniques for micro-to-nanoscale devices, such as diatom-based energy producers. Sandhage (2006) highlighted the reactive conversion of biologically produced micro templates that had previously been shown to function as gas sensors and may be used in energy-generating devices. Sandhage & Bao (2009) then investigated the manufacture of nano- to microscale diatom frustule structures for use as diodes or P–N junctions in silicon-based solar cells, or as high-surface-area battery electrodes in lithium batteries, among other applications.

8.4 FUTURE PROSPECTS OF DIATOM RESEARCH

The documented studies stated diatoms' amazing nature and potential all-around use. For example, advances in diatom research can lead to novel products in the domains of drug delivery, sensing, and detection components for complex biomedical devices, as well as nanoparticles for waste degradation. Furthermore, recent advances in sequencing technology and the processing of large biological datasets have enabled the labelling and storage of the global biodiversity of diatoms in all geographical locations.

Scaling up the process for large-scale manufacture, which is dependent on various micro and macro aspects such as cultivation, harvesting, drying, genetic manipulation, dearth of genomic, proteomic, and metabolic data, etc., is one of the primary hurdles in the diatom-based industry. However, with improvements in genetic tools, bioreactors, as well as other infrastructure upgrades, it may be able to overcome these difficulties in the near future. In general, there are numerous obstacles in the way of bio-based companies at all levels, including academic and industrial research, infrastructure, laws, and informational gaps. The papers show that scientific work and discoveries are progressing, but other areas must also contribute, such as infrastructure development, closing the knowledge gap between scientists and businesspeople, and altering both national and international policy.

To increase the rate of innovation in the biomedical business without jeopardizing the screening procedure defined by public health authorities, an original and distinctive method is needed to attract academic researchers and bio-entrepreneurs. Entrepreneurs and researchers working together will enable a complete assessment of the market for scientific breakthroughs, manufacturing, investment, and product globalization. Given the global biomedical infrastructure's rapid developments, it seems possible. The fast developments in response to COVID-19 have shown this, and they should be embraced to be used in other biotech-based sectors (Harris et al., 2020). In the near future, it is feasible that advancements in the use of silica-based support structures for drug delivery will take place concurrently with changes in hospital infrastructure and the pharmaceutical sector. The other prerequisite is to bridge the gap between entrepreneurs and academic researchers in terms of the huge and intricate scientific information and knowledge.

8.5 CONCLUSION

Diatoms have played a critical role in the ecosystem for millions of years, serving as one of the most important sources of oxygen synthesisers on the planet and as a key supplier of biomass in the seas besides other groups of algae. The diatom-produced silica templates are inexpensive, readily repeatable, durable, and mechanically strong, and their development requires just basic nutrients and light. On a general note, the frustule defines a cell and regulates the movement of chemicals inside that cell. It is the most important biological barrier in the function of diatoms. Silica is used to make the frustule stiff and precise, which serves as a barrier. Diatoms are an excellent model organism for studying micro and nanoscale contortions because their geometric regularity makes changes visible, their production of 20% of global organic carbon fixation makes them important in the biosphere, and their widespread distribution from sunlit soils to the open ocean makes the principles learned applicable across most of the biosphere, as well as in chemistry, biology, and engineering.

In the end, it can be said that diatoms – the gifts of Mother Nature – are very tiny nanosilica-encapsulated cells with intricate taxonomic ultrastructural patterns. Moreover, with unique nanostructures, diatom frustules can be modulated in different ways for exploitation in several nano-biotechnological fields.

REFERENCES

Armbrust E. V. (2009) The life of diatoms in the world's oceans. *Nature* 459: 185–192.

Bao Z., Weatherspoon M. R., Shian S., Cai Y., Graham P. D., Allan S. M., Ahmad G., Dickerson M. B., Church B. C., Kang Z., Abernathy Iii H. W., Sandhage K. H. (2007) Chemical reduction of three-dimensional silica micro-assemblies into microporous silicon replicas. *Nature* 446(7132): 172–175.

Belegratis M., Schmidt V., Nees D., Stadlober B., Hartmann P. (2013) Diatom-inspired templates for 3D replication: Natural diatoms *versus* laser written artificial diatoms. *Bioinspiration Biomimetics* 9(1): 016004.

Bismuto A., Setaro A., Maddalena P., De Stefano L., De Stefano M. (2008) Marine diatoms as optical chemical sensors: A time-resolved study. *Sens Actuators B: Chem* 130: 396–399.

Bondareva J. V., Aslyamov T. F., Kvashnin A. G., Dyakonov P. V., Kuzminova Y. O., Mankelevich Y. A., Voronina E. N., Dagesyan S. A., Egorov A. V., Khmelnitsky R. A., Tarkhov M. A., Suetin N. V., Akhatov I. S., Evlashin S. A. (2020) Environmentally friendly method of silicon recycling: Synthesis of silica nanoparticles in an aqueous solution. *ACS Sustainable Chem Eng* 8: 14006–14012.

Bose R., Roychoudhury P., Pal R. (2021) In-situ green synthesis of fluorescent silica–silver conjugate nanodendrites using nanoporous frustules of diatoms: An unprecedented approach. *Bioprocess Biosyst Eng* 44(6): 1263–1273.

Bose R., Saha R., Chattopadhyay S., Pal R. (2020) Utilization of nanoporous biosilica of diatoms as a potential source material for fabrication of nanoelectronic device and their characterization. *J Appl Phycol* 32: 3041–3049.

Bradbury J. (2004) Nature's nanotechnologists: unveiling the secrets of diatoms. *PLoS Biol* 2: 1512–1515.

Breslin M., Ringnalda J., Xu L., Fuller M., Seeger J., Daehn G., Otani T., Fraser H. (1995) Processing, microstructure, and properties of co-continuous alumina-aluminum composites. *Mater Sci Eng A* 195: 113–119.

Brown M. R. (1991) The amino-acid and sugar composition of 16 species of microalgae used in mariculture. *J Exp Mar Biol Ecol* 145: 79–99.

Brown M. R., Miller K. A. (1992) The ascorbic acid content of eleven species of microalgae. *J Appl Phycol* 4: 205–215.

Brown M. R., Mular M., Miller I., Farmer C., Trenerry C. (1999) The vitamin content of microalgae in aquaculture. *J Appl Phycol* 11: 247–255.

Butcher K. S. A., Ferris J. M., Phillips M. R. (2003) Photoluminescence and cathodoluminescence studies of diatoms – nature's own nano-porous silica structures. 27th Annual Condensed Matter and Materials Meeting, 4th–7th February 2003, In: D. Paganin, A. Smith and G. Troup (eds.) *Proceedings of the 27th Annual A&ZNIP Condensed Matter and Materials Meeting*, Australian and New Zealand Institute of Physics.

Cavalier-Smith T., Chao E. E. (1996) 18S rRNA sequence of *Heterosigma carterae* (Raphidophyceae), and the phylogeny of heterokont algae (Ochrophyta). *Phycologia* 35(6): 500–510.

Cox E. J. (2015) Diatoms, diatomeae (Bacillariophyceae s.l., Bacillariophyta). In: Syllabus of Plant Families (Ed. by W. Frey), Borntraeger, Stuttgart. pp. 64–103.

Crawford R. M., Gardner C.,Medlin L. K. (1994) The genus *Attheya*. I. A description of four new taxa, and the transfer of *Gonioceros septentrionalis* and *G. armata*. *Diat Res* 9: 27–51.

De Stefano L., De Stefano M., De Tommasi E., Rea I., Rendina I. (2011) A natural source of porous biosilica for nanotech applications: The diatoms microalgae. *Physica Status Solidi C* 8(6): 1820–1825.

De Stefano L., Maddalena P., Moretti L., Rea I., Rendina I., De Tommasi E., Mocella V., De Stefano M. (2009) Nano-biosilica from marine diatoms: A brand new material for photonic applications. *Superlattice Microst* 46: 84–89.

De Stefano L., Rendina I., De Stefano M., Bismuto A., Maddalena P. (2005) Marine diatoms as optical chemical sensors. *Appl Phys Lett* 87: 233902.

Dempsey E., Diamond D., Smyth M. R., Urban G., Jobst G., Moser I., Verpoorte E. M., Manz A., Widmer H. M., Rabenstein, K., Freaney R. (1997) Design and development of a miniaturised total chemical analysis system for on-line lactate and glucose monitoring in biological samples. *Anal Chim Acta* 346: 341–349.

Dickerson M., Sandhage K., Nalik R., Stone M. (2007) Methods for fabricating micro-to-nanoscale devices via biologically-induced solid formation on biologically-derived templates, and micro-to-nanoscale structures and micro-to-nanoscale devices made thereby. U.S. Patent Application 11/358,342.

Dunstan G. A., Volkman J. K., Barrett S. M., Leroi J. M., Jeffrey S. W. (1993) Essential polyunsaturated fatty acids from 14 species of diatom (Bacillariophyceae). *Phytochemistry* 35 (1): 155–161.

Fang Y., Wu Q., Dickerson M. B., Cai Y., Shian S., Berrigan J. D., Poulsen N., Kroger N., Sandhage K. H. (2009) Protein-mediated layer-by-layer syntheses of freestanding microscale titania structures with biologically assembled 3-D morphologies. *Chem Mater* 21: 5704–5710.

Farjadian F., Roointan A., Mohammadi-Samani S., Hosseini M. (2019) Mesoporous silica nanoparticles: Synthesis, pharmaceutical applications, biodistribution, and biosafety assessment. *Chem Eng J* 359: 684–705.

Gale D. K., Gutu T., Jiao J., Chang C. H., Rorrer G. L. (2009) Photoluminescence detection of biomolecules by antibody-functionalized diatom biosilica. *Adv Funct Mater* 19(6): 926–933.

Gammone M. A., D'Orazio N. (2015) Anti-obesity activity of the marine carotenoid fucoxanthin. *Mar Drugs* 13 (4): 2196–2214.

Geitler L. (1932) Der Formwechsel der pennaten Diatomeen (Kieselalgen). *Archiv für Protistenkunde* 78: 1–226.

Ghotekar S. (2019) A review on plant extract mediated biogenic synthesis of CdO nanoparticles and their recent applications. *Asian J Green Chem* 3: 187–200.

Gordon R., Sterrenburg F. A. S., Sandhage K. H. (2005) A special issue on diatom nanotechnology. *J Nanosci Nanotech* 5: 1–4.

Gupta S., Kashyap M., Kumar V., Jain P., Vinayak V., Joshi K. B. (2018) Peptide mediated facile fabrication of silver nanoparticles over living diatom surface and its application. *J Mol Liq* 249: 600–608.

Gutu T., Gale D. K., Jeffryes C., Wang W., Chang C. H., Rorrer G. L., Jiao J. (2009) Electron microscopy and optical characterization of cadmium sulphide nanocrystals deposited on the patterned surface of diatom biosilica. *J Nanomater* 2009: 1–7. 10.1155/2009/860536.

Hamm C. E., Merkel R., Springer O., Jurkojc P., Maier C., Prechtel K., Smetacek V. (2003) Architecture and material properties of diatom shells provide effective mechanical protection. *Nature* 421(6925): 841–843.

Harwood D. M. (1999) Diatomite. In: *The Diatoms: Applications for the Environmental and Earth Sciences* (Eds. by E. F. Stoermer and J. P. Smol), Cambridge University Press, Cambridge. pp. 436–446.

Harris M., Bhatti Y., Buckley J., Sharma D. (2020) Fast and frugal innovations in response to the COVID-19 pandemic. *Nat Med* 26: 814–817.

Hasle G. R., Syvertsen E. E. (1997) Marine diatoms. In: *Identifying marine phytoplankton* (Ed. by C. R. Tomas), Academic Press, San Diego. pp. 5–385.

Hoek, C., Mann, D., Jahns, H. M., Jahns, M. (1995) *Algae: an introduction to phycology*. Cambridge university press.

Hustedt F. (1927–1966) Die Kieselalgen Deutschlands. *Österreichs und der Schweiz*. 3 vols. Akademische Verlagsgesellschaft, Leipzig.

Iravani S. (2011) Green synthesis of metal nanoparticles using plants. *Green Chem* 13: 2638–2650.

Iravani S., Jamalipour S. G. (2019) Gold Nanostructures in Medicine and Biology. In *Nanoparticles in Medicine* (Ed. by A. Shukla), Springer, Singapore.

Iravani S., Varma R. (2019) Plant-derived edible nanoparticles and miRNAs: Emerging frontier for therapeutics and targeted drug delivery. *ACS Sustainable Chem Eng* 7: 8055–8069.

Jeffryes C., Gutu T., Jiao J., Rorrer G. L. (2008) Metabolic insertion of nanostructured TiO_2 into the patterned biosilica of the diatom *Pinnularia* sp. by a two-stage bioreactor cultivation process. *Acs Nano* 2(10): 2103–2112.

Kato K., Lee S., Nagata F. (2020) Efficient enzyme encapsulation inside solgel silica sheets prepared by poly-L-lysine as a catalyst. *J Asian Ceramic Soc* 8: 396–406.

Kaehr B., Shear J. B. (2007) Mask-directed multiphoton lithography. *J Am Chem Soc* 129(7): 1904–1905.

Karsten G. (1928) Bacillariophyta (Diatomaceae). In *Die Natürlichen Pflanzenfamilien* (Ed.by A. Engler and K. Prantl) (Vol. 2, 2nd ed.). W. Engelmann, Leipzig. pp. 105–203.

Khripin C. Y., Pristinski D., Dunphy D. R., Brinker C. J., Kaehr B. (2011) Protein-directed assembly of arbitrary three-dimensional nanoporous silica architectures. *ACS Nano* 5(2): 1401–1409.

Kiani M., Rabiee N., Bagherzadeh M., Ghadiri A. M., Fatahi Y., Dinarvand R., Webster T. J. (2021) Improved green biosynthesis of chitosan decorated Ag-and Co3O4-nanoparticles: A relationship between surface morphology, Photocatalytic and biomedical applications. *Nanomedicine* 32: 102331.

Kiani M., Rabiee N., Bagherzadeh M., Ghadiri A. M., Fatahi Y., Dinarvand R., Webster T. J. (2020) High-gravity-assisted green synthesis of palladium nanoparticles: The flowering of nanomedicine. *Nanomedicine* 30: 102297.

Kooistra H., Gersonde R., Medlin L., Mann D. (2007) The origin and evolution of the diatoms: their adaptation to a planktonic existence. In *Evolution of Primary Producers in the Sea* (Eds. by P. G. Falkowski, and A. Knoll), Elsevier Academic Press, Amsterdam. pp. 207–249.

Krammer K., Lange-Bertalot H. (1986) Bacillariophyceae 1. Teil: Naviculaceae. In *Susswasserflora von Mitteleuropa* (Eds. by H. Ettl, J. Gerloff, H. Heynig and D. Mollenhauer) 2(l), G. Fischer, Stuttgart & New York. p. 876.

Krammer K., Lange-Bertalot H. (1988) Bacillariophyceae 2. Teil: Bacillariaceae, Epithemiaceae, Surirellaceae. In *Susswasserflora von Mitteleuropa* (Eds. by H. Ettl, J. Gerloff, H. Heynig and D. Mollenhauer) 2(2), G. Fischer, Stuttgart C New York. p. 596.

Krammer K., Lange-Bertalot H. (1991a) Bacillariophyceae 3. Teil: Centrales, Fragilariaceae, Eunotiaceae. In *Susswasserflora von Mitteleuropa* (Eds. by H. Ettl, J. Gerloff, H. Heynig and D. Mollenhauer) 2(3), G. Fischer, Stuttgart & New York. p. 576.

Krammer K., Lange-Bertalot H. (1991b) Bacillariophyceae 4. Teil: Achnanthaceae K. ritische Ergiinzungenz u *Navicula* (Lineolutae) and *Gomphonema.* In *Susswasserflora von Mitteleuropa* (Eds. by H. Ettl, G. Gartner, J. Gerloff, H. Heynig and D. Mollenhauer) 2(4), G. Fischer, Stuttgart & New York. p. 437.

Krammer K., Lange-Bertalot H. (2000) Bacillariophyceae 3. Centrales, Fragilariaceae, Eunotiaceae. Erganzte und korrigierte 2. Auflage, In: *Suswasserflora von Mitteleuropa* (Eds. by H. Ettl, J. Gerloff, H. Heynig and D. Mollenhauer) vols. 2/3, Spektrum Akademischer Verlag, Heidelberg/Berlin, pp. 1–598.

Krammer K., Lange-Bertalot H. (2004) Bacillariophyceae 4. Achnanthaceae, Kritische Erganzungen zu *Navicula* (Lineolatae) und *Gomphonema* Gesamtliteraturverzeichnis Teil 1–4. 2. Auflage. In *Suswasserflora von Mitteleuropa* (Eds. by H. Ettl, J. Gerloff, H. Heynig and D. Mollenhauer) vols. 2/4, Spektrum Akademischer Verlag, Heidelberg/Berlin, pp. 1–468.

Kutzing F. T. (1834a) Synopsis diatomearum oder Versuch einer systematischen Zusammenstellung der Diatomeen. *Linnaea* 8: 529–629.

Kutzing F. T. (1834b) *Synopsis Diatomearum oder, Versuch einer systematischen Zusammenstellung der Diatomeen.* Schwetschke, Halle.

Kützing F. T. (1844) *Die kieselschaligen Bacillarien oder Diatomeen.* W. Köhne, Nordhausen.

Lang Y., Monte F. D., Rodriguez B. J., Dockery P., Finn D. P., Pandit A. (2013) Integration of TiO_2 into the diatom *Thalassiosira weissflogii* during frustule synthesis. *Sci Rep* 3(1): 1–11.

Lasagni A., Menéndez-Ormaza B. (2010) Two-and three-dimensional micro-and sub-micrometer periodic structures using two-beam laser interference lithography. *Adv Eng Mater* 12(1–2): 54–60.

Lechner C. C., Becker F. W. C. (2015) Silaffins in silica biomineralization and biomimetic silica precipitation. *Mar Drugs* 13: 5297–5333.

Levkov Z. (2009) *Amphora* sensu lato. In: *Diatoms of Europe: Diatoms of the European Inland Waters and Comparable Habitats* (Ed. by H. Lange-Bertalot) Vol. 5, A.R.G. Gantner Verlag K.G., 5: 5-916.

Lombardi A., Wangersky P. J. (1995) Particulate lipid class composition of three marine phytoplankters *Chaetoceros gracilis*, *Isochrysis galbana* (Tahiti) and *Dunaliella tertiolecta* grown in batch culture. *Hydrobiologia* 306: 1–6.

Losic D., Mitchell J. G., Voelcker N. H. (2005) Complex gold nanostructures derived by templating from diatom frustules. *Chem Commun (Camb)* 39: 4905–4907.

Losic D., Mitchell J. G., Lal R., Voelcker N. H. (2007) Rapid fabrication of micro and nanoscale patterns by replica molding from diatom biosilica. *Adv Funct Mater* 17: 2439–2446.

Losic D., Mitchell J. G., Voelcker N. H. (2009) Diatomaceous lessons in nanotechnology and advanced materials. *Adv Mater* 21: 2947–2958.

Lund J. W. G. (1965) The ecology of the freshwater phytoplankton. *Biol Rev* 40: 231–293.

Lund J. W. G., Reynolds C. S. (1982) The development and operation of large limnetic enclosures in Blelham Tarn, English Lake District, and their contribution to phytoplankton ecology. *Progr Phycol Res* 1: 1–65.

Machill S., Kohler L., Ueberlein S., Hedrich R., Kunaschk M., Paasch S., Schulze R., Brunner E. (2013) Analytical studies on the incorporation of aluminium in the cell walls of the marine diatom *Stephanopyxis turris*. *BioMetals* 26: 141–150.

Mangas-Sánchez J., Adlercreutz P. (2015) Highly efficient enzymatic biodiesel production promoted by particle-induced emulsification. *Biotechnol Biofuels* 8(1): 1–8.

Mann D. G. (1999) The species concept in diatoms. *Phycologia* 38(6): 437–495.

Medlin L., Kooistra WHCF, Schmid A. M. (2000) A review of the evolution of the diatoms-a total approach using molecules, morphology and geology. The origin and early evolution of the diatoms: fossil, molecular and biogeographical approaches. In *Polish Academy of Sciences* (Eds. by A. Witkowski and J. Sieminska), Krakow, Poland. pp. 13–35.

Medlin L. K., Kaczmarska I. (2004) Evolution of the diatoms: V. Morphological and cytological support for the major clades and a taxonomic revision. *Phycologia* 43: 245–270.

Metzeltin D., Lange-Bertalot H. (2007) Tropical diatoms of South America II. Special remarks on biogeographic disjunction. In *Iconographia diatomologica. Annotated diatom micrographs, Diversity–taxonomy–biogeography* (Ed. by H. Lange-Bertalot) (Vol. 18), A.R.G. Gantner, Ruggell

Meyer H., Drewer H., Gruendig B., Cammann K., Kakerow R., Manoli Y., Mokwa W., Rospert M. (1995) Two-dimensional imaging of O_2, H_2O_2, and glucose distributions by an array of 400 individually addressable microelectrodes. *Anal Chem* 67 (7): 1164–1170.

Mishra B., Saxena A., Tiwari A. (2020) Biosynthesis of silver nanoparticles from marine diatoms *Chaetoceros* sp., *Skeletonema* sp., *Thalassiosira* sp., and their antibacterial study. *Biotechnol Rep* 28: e00571.

Morel F. M. M., Price N. M. (2003) The biogeochemical cycles of trace elements in the oceans. *Science* 300: 944–947.

Moretti L., De Stefano L., Rendina I. (2007a) Quantitative analysis of capillary condensation in fractal-like porous silicon nanostructures. *J Appl Phys* 101(2): 024309.

Moretti L., Rea I., De Stefano L., Rendina I. (2007b) Periodic versus aperiodic: Enhancing the sensitivity of porous silicon based optical sensors. *Appl Phys Lett* 90(19): 191112.

Nelson D. M., Tréguer P., Brzezinski M. A., Leynaert A., Quéguiner B. (1995) Production and dissolution of biogenic silica in the ocean: Revised global estimates, comparison with regional data and relationship to biogenic sedimentation. *Glob Biogeochem Cycles* 9(3): 359–372.

Nielson R., Kaehr B., Shear J. B. (2009) Microreplication and design of biological architectures using dynamic-mask multiphoton lithography. *Small* 5(1): 120–125.

Ongerth J. E., Hutton A. P. (1997) DE filtration to remove Cryptosporidium. *J Am Water Works Assoc* 89(12): 39–46.

Osman S., Gören R., Özgür C. (2009) Purification of diatomite powder by acid leaching for use in fabrication of porous ceramics. *Int J Miner Proc* 93(1): 6–10.

Paasche E. (1980) Silicon. In *The Physiological Ecology of Phytoplankton. Studies in Ecology 7* (Ed. by I. Morris), Blackwells, Oxford. pp. 259–284.

Pamirsky I. E., Golokhvast K. S. (2013) Silaffins of diatoms: From applied biotechnology to biomedicine. *Mar Drugs* 11: 3155–3167.

Parkinson J., Gordon R. (1999) Beyond micromachining: The potential of diatoms. *Trends Biotechnol* 17: 190–196.

Pfitzer E. (1871) *Untersuchungen uber Bau und Entwickelung der Bacillariaceen (Diatomaceen)*. Bonn, Adolf Marcus [as Heft 2 of J. Hanstein, *Botanische Abhandlungen aus dem Gebiet der Morphologie und Physiologie*. 1872].

Pickett-Heaps J., Schmid A-MM, Edgar L. A. (1990) The cell biology of diatom valve formation. *Progr Phycol Res* 7: 1–168.

Pytlik N., Kaden J., Finger M., Naumann J., Wanke S., Machill S., Brunner E. (2017) Biological synthesis of gold nanoparticles by the diatom *Stephanopyxis turris* and in vivo SERS analyses. *Algal Res* 28: 9–15.

Qin T., Gutu T., Jiao J., Chang C. H., Rorrer G. L. (2008) Biological fabrication of photoluminescent nanocomb structures by metabolic incorporation of germanium into the biosilica of the diatom *Nitzschia frustulum. Acs Nano* 2 (6): 1296–1304.

Reynolds C. S. (1984) *The ecology of freshwater phytoplankton*. Cambridge University Press, Cambridge, p. 384.

Rorrer G. L., Chang C. H., Liu S. H., Jeffryes C., Jiao J., Hedberg J. A. (2005) Biosynthesis of silicon-germanium oxide nanocomposites by the marine diatoms *Nitzschia frustulum. J Nanosci Nanotechnol* 5: 41–49.

Round F. E. (1981) *The ecology of algae*. Cambridge University Press, USA, pp. 664.

Round F. E., Crawford R. M., Mann D. G. (1990) *Diatoms: biology and morphology of the genera*. Cambridge University Press, pp 747.

Roychoudhury P., Golubeva A., Dąbek P., Gloc M., Dobrucka R., Kurzydłowski K., Witkowski A. (2021) Diatom mediated production of fluorescent flower shaped silver-silica nanohybrid. *Materials* 14(23): 7284.

Roychoudhury P., Nandi C., Pal R. (2016) Diatom-based biosynthesis of gold-silica nanocomposite and their DNA binding affinity. *J Appl Phycol* 28: 2857–2863.

Sachindra N. M., Sato E., Maeda H., Hosokawa M., Niwano Y., Kohno M., Miyashita K. (2007) Radical scavenging and singlet oxygen quenching activity of marine carotenoid fucoxanthin and its metabolites. *J Agric Food Chem* 55 (21): 8516–8522.

Sandhage K. H. (2006) Shaped microcomponent via reactive conversion of biologically-derived microtemplates. U.S. Patent 7,067,104.

Sandhage K. H., Bao Z. (2009) Methods of fabricating nanoscale-to-microscale structures. U.S. Patent 7,615,206.

Sandhage K. H., Dickerson M. B., Huseman P. M., Caranna M. A., Clifton J. D., Bull T. A., Heibel T. J., Overton W. R., Schoenwaelder M. E. A. (2002) Novel, bioclastic route to self-assembled, 3D, chemically tailored meso/nanostructures: Shape-preserving reactive conversion of biosilica (diatom) microshells. *Adv Mater* 14: 429–433.

Schmidt A. (1874–1959) Atlas der Diatomaceen-kunde. *Aschersleben, Commissions-Verlag Von Ludwig Siever's Buchandlung, O.R. Reisland.* Akademie-Verlag, Leipzig, Berlin.

Schröfel A., Kratošová G., Bohunická M., Dobročka E., Vávra I. (2011) Biosynthesis of gold nanoparticles using diatoms—silica gold and EPS-gold bionanocomposite formation. *J Nanopart Res* 13: 3207–3216.

Schuler P. F., Ghosh M. M., Gopalan P. (1991) Slow sand and diatomaceous earth filtration of cysts and other particulates. *Water Res* 25 (8): 995–1005.

Schutt F. (1896) Bacillariales. In *Die Natürlichen Pflanzenfamilien* (Eds. by A. Engler and K. Prantl), vol. 1(1b), W. Englemann, Leipzig. pp. 31–153.

Seckbach J., Kociolek P. (Eds.). (2011) *The diatom world* (Vol. 19). Springer Science & Business Media.

Silva P. C. (1962) Classification of algae. In *Physiology and biochemistry of algae* (Ed. by R. A. Lewin), Academic, New York. pp. 827–840.

Sims P. A., Mann D. G., Medlin L. K. (2006) Evolution of the diatoms: Insights from fossil, biological and molecular data. *Phycologia* 45: 361–402.

Stewart M. P., Buriak J. M. (2000) Chemical and biological applications ofporous silicon technology. *Adv Mater* 12(12): 859–869.

Sumper M., Brunner E. (2006) Learning from diatoms: Nature's tools for the production of nanostructured silica. *Adv Funct Mater* 16: 17–26.

Sumper M., Kroger N. (2004) Silica formation in diatoms: The function of long-chain polyamines and silaffins. *J Mater Chem* 14: 2059–2065.

Szczech J. R., Jin S. (2008) Mg_2Si nanocomposite converted from diatomaceous earth as a potential thermoelectric nanomaterial. *J Solid State Chem* 181(7): 1565–1570.

Tam L. T., Van Cong N., Ha N. C., Hang N. T. M., Van Minh C., Hong D. D. (2021) Cultivation and biomass production of the diatom *Thalassiosira weissflogii* as a live feed for white-leg shrimp in hatcheries and commercial farms in Vietnam. *J Appl Phycol* 33(3): 1559–1577.

Tavakolizadeh M., Pourjavadi A., Ansari M., Tebyanian H., Seyyed Tabaei S. J., Atarod M., Rabiee N., Bagherzadeh M., Varma R. S. (2021) An environmentally friendly wound dressing based on a self healing, extensible and compressible antibacterial hydrogel. *Green Chem* 23: 1312–1329.

Townley H. E., Parker A. R., White-Cooper H. (2008) Exploitation of diatom frustules for nanotechnology: Tethering active biomolecules. *Adv Funct Mater* 18(2): 369–374.

Townley H. E., Woon K. L., Payne F. P., White-Cooper H., Parker A. R. (2007) Modification of the physical and optical properties of the frustule of the diatom *Coscinodiscus wailesii* by nickel sulfate. *Nanotechnology* 18: 295101.

Van Den Hoek C., Mann D. G., Jahns H. M. (1995) *Algae: Introduction to phycology*. Cambridge University Press, Cambridge, UK, p. 623.

Varma R. S. (2012) Greener approach to nanomaterials and their sustainable applications. *Curr Opin Chem Eng* 1: 123–128.

Varma R. S. (2019) Biomass-derived renewable carbonaceous materials for sustainable chemical and environmental applications. *ACS Sustainable Chem Eng* 7: 6458–6470.

Villareal T. A., Pilskaln C., Brzezinski M., Lipschultz F., Dennett M., Gardner G. B. (1999) Upward transport of oceanic nitrate by migrating diatom mats. *Nature* 397: 423–425.

Wetherbee R., Crawford S., Mulvaney P. (2000) The nanostructure and development of diatom biosilica. In *Biomineralization: from biology to biotechnology and medical application* (Ed. by E. Baeuerlein), Wiley-VCH, Weinheim, pp 189–206.

Yamazaki, T., Sasanuma, H., Mayama, S., Umemura, K. (2010). Cultivation of Melosira nummuloides cells in the presence of platinum: Preparation of metal-containing frustules. *physica status solidi c* 7(11-12): 2759–2762.

Yang J., Zhen L., Ren F., Campbell J., Rorrer G. L., Wang A. X. (2015) Ultra-sensitive immunoassay biosensors using hybrid plasmonic-biosilica nanostructured materials. *J Biophotonics* 8(8): 659–667.

Zamora P., Narváez A., Domínguez E. (2009) Enzyme-modified nanoparticles using biomimetically synthesized silica. *Bioelectrochemistry* 76: 100–106.

Zhang H., Tang Y., Zhang Y., Zhang S., Qu J., Wang X., Kong R., Han C., Liu Z. (2015) Fucoxanthin: A promising medicinal and nutritional ingredient. *Evid Based Complement Alternat Med* 2015: 1–10.

9 Cultivation of Diatoms in Photobioreactors

C. Y. Tong and C. J. C. Derek
School of Chemical Engineering, Engineering Campus, Universiti Sains Malaysia Nibong Tebal, Penang, Malaysia

CONTENTS

9.1 INTRODUCTION

Microalgae are the unicellular ancestors to terrestrial plants with the capabilities of utilizing light energy for synthesis of organic materials from carbon dioxide, water, and nutrients via photosynthesis (Madigan et al., 2006). They are deemed as the earliest life-forms on earth with the absence of leaves, roots, or stems and a very simple cell structure, hence rendering them quicker access to CO_2 and nutrients than terrestrial plants (Mantzorou & Ververidis, 2019). Being a vast group of unicellular organisms comprising eukaryotic protists, prokaryotic cyanobacteria, and blue-green algae, microalgae are well known as oxygen producers through photoautotrophic metabolic pathways. Under heterotrophic cultivation, microalgae would require a continuous supply of oxygen to metabolize the organic carbon sources supplied. Besides photoautotrophic and heterotrophic cultivation mode, microalgae could be even cultured mixotrophically with both light and organics/inorganic carbon as energy sources. Most of the microalgae are capable of adapting themselves to wide ranges of environments and to a certain extent to moving from freshwater to seawater and vice versa. The marine microalgae denote a larger potential contributor over freshwater microalgae in terms of carbon sink due to the significantly large land area that is covered by

DOI: 10.1201/9781003322115-9

the sea (Merlo et al., 2021). Some microalgae, such as diatoms, have a much higher nutrient uptake rate than the other algae classes, rendering them as good nutrient sequesters (Litchman, 2007).

Diatoms, often unicellular microalgae (typically 10–200 μm in size), are distinguished by the presence of a transparent siliceous frustule with two valves and a separating girdle band; they are able to grow and colonize submerged surfaces rapidly in every hospitable aquatic environment (Cameron, 2013). They are usually referred to as non-motile golden-brown eukaryotes with limited mobility along the substrate via the release of mucilaginous material along a slit-like grove called raphe. Typically, diatoms are classified into the division Chrysophyta, class of Bacillariophyceae. They are always found abundantly in micro-colonies, attaching by mucous filaments, and are responsible for 20% of global carbon fixation in the ocean (Hildebrand et al., 2012; Jones, 2013). Diatoms can be further classified into centric diatoms (Coscinodiscophyceae), pennate diatoms without a raphe (Fragilariophyceae), and pennate diatoms with raphe (Bacillariophyceae) by existing in colonies, filaments, ribbons (Fragilaria), fans (Meridion), zigzags (Tabellaria), or stellate (Asterionella) (Mishra et al., 2017). They can inhabit both marine (~75%) and freshwater or other habitats (~25%) with more than 10,000 species, appearing as either phytoplankton (suspended in the water column) or attached to substratum (benthos) (Jain, 2020). Typically, pennate diatoms dominate the biofilm community, whereas centric diatoms dominate the water column (Patil & Anil, 2005). Diatoms are easily acquired and proliferate rapidly via size reduction division at each generation until reaching a minimal size.

These primitive single cells with large surface-to-volume body ratio have several advantages: (1) short doubling time and no seasonal limitations with some species able to survive in extremely harsh environments, (2) high capability for year-round mass production under well-controlled process parameters, (3) no competition with terrestrial plants for agricultural land, (4) minimum environmental issues due to minimal fertilizers and no pesticides requirement, and (5) high accessibility to the nutrients from wastewater. Their growth can be easily regulated by adjusting the silicate concentration, as diatom bio-silicate walls only require a small amount of energy to be synthesized (around 8% of a comparable organic cell wall), allowing them to grow faster by channeling the overall cell energy elsewhere (Wang & Seibert, 2017). With lipid fractions ranging from 15% to 85% and highly valued products with various applications, including food, cosmetics, aquaculture feed, fertilizers, and biopharmaceutical, diatoms will definitely gain extensive attention for mass production in the future to cater to the needs of commercial markets (Cheah & Chan, 2021; Cheah et al., 2022; Wang & Seibert, 2017).

9.2 CONVENTIONAL CULTIVATION MODE OF DIATOMS

In order to meet the escalating demands of microalgal biomass, the two most conventional cultivation methods are open systems and controlled closed systems, as illustrated in Figure 9.1. Major open systems are large open raceway ponds, circular ponds, and unstirred ponds. Unstirred ponds appear to be the least productive among the three, though they are the least expensive to operate, as absence of a mixing apparatus in the system reduces microalgal growth, particularly in suspension mode. Over time, aged microalgal cells tend to lose their mobility and acquire lower electrostatic repulsion between themselves, which leads to cell clumping and eventually mass sedimentation when the cells settle at the bottom of the pond. Circular ponds with a center pivot mixer showed better biomass productivities, but unfortunately, they were limited by the pond size due to the engineering constraints on the pivot mixer (Liang et al., 2015). Raceway ponds, on the other hand, were the most widely used open algal cultivation technique in the world.

Open raceway pond is a shallow raceway about 0.15 to 0.45 m in depth and constructed with a concrete, clay, or plastic-lined pond with lower capital costs of construction. Typically, it is accompanied by paddle wheels to ensure uniform mixing of the microalgal suspension and facilitates both the gas and nutrient distribution in the suspension. Aside from that, numerous baffles are installed to guide the flow around the bends. Such a system is considered to be the first

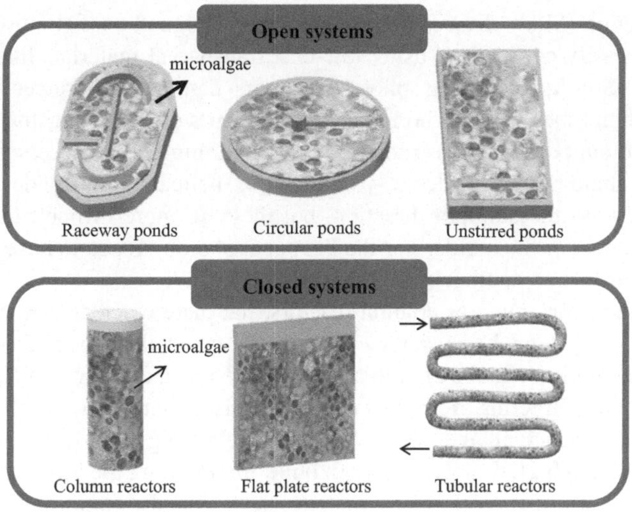

FIGURE 9.1 Open and closed cultivation systems of microalgae (adapted and modified from (Djamal & Henni, 2019)).

attempt to mass-cultivate microalgae. Over the years, artificial man-made outdoor raceway ponds were adapted in industrial production scale (Casini et al., 2011). An outdoor mass cultivation of an oleaginous diatom, *Fistulifera solaris*, was established for year-round synthesis of microalgal lipids; nevertheless, the authors encountered a bottleneck in the cultivation during the winter since this type of diatom was difficult to sustain in cold conditions (<15°C) (Sato et al., 2014). Therefore, they intentionally selected a cold-tolerant diatom, *Mayamaea* sp., for biofuel production during winter, and it has demonstrated successful areal productivities of biomass and lipids up to 8.62 and 0.92 g m^{-2} day,$^{-1}$ respectively (Matsumoto et al., 2017).

The efficiency of an open pond system is significantly influenced by local diurnal temperature variations and solar insolation fluctuations. They are easy to be operated but concurrently accompanied by low volumetric productivities and high contamination risk of the culture from bacterial loads and grazers (Narala et al., 2016). Since the open pond systems are constantly exposed to the ambient air, water evaporation is going to be a huge concern; therefore, the pond depth should also be balanced between the need to maintain an adequate water depth for mixing and the need to avoid large changes in ionic strength or light irradiation limitation (Sanz Smachetti et al., 2018). It is a daunting challenge to get proper control of the nutrient availability and light intensity due to the vagaries of weather. Furthermore, since open pond cultivation has lower controllability, it is more suitable for extremophiles such as *Spirulina* sp. (high pH) and *Dunaliella* sp. (high salinity), which are capable of suppressing other algal species or predators (Hoysall et al., 2012). Otherwise, the contaminants might outcompete the desired algal species while the predators or rotifers would decimate the algal culture (Xu et al., 2009).

Closed photobioreactors, on the other hand, facilitate higher areal biomass productivities. For instance, bubble column or air-lift reactors are simple gas-liquid contacting devices that gas is injected into the liquid from the column bottom by an appropriate distributor. The gas can be utilized to agitate the microalgal cultures within the column and promote medium circulation in the reactors (Sato et al., 2014). Other examples are flat panel reactors, a rectangular airlift photobioreactor consisting of a large number of light-distributing plates or panels fixed a few centimeters away from each other, with mixing provided by air injected between adjacent plates. These panels have a large, illuminated surface area to culture volume that keeps the light uniformly distributed across the entire reactor vessel, and they are also simple in structure. They are preferred to be illuminated from one side by direct sunlight and vertically positioned in east-west orientation for microalgae proliferation (Show & Lee, 2014). Due to the designed configuration, the material cost is moderate as it is easily constructed using low-cost transparent materials like glass, plexi-glass, polycarbonate, etc. Similarly, tubular photobioreactors can be constructed using inexpensive polyvinyl chloride film tubes with the horizontal tubular parts of the bioreactors laid on the ground without complicated supporting infrastructure, enabling commercial success for microalgae cultivation. In order to attain higher surface-to-volume ratios, helical or conical designs are applied for better angling relative to the sunlight direction, but they are more difficult to scale up (Burgess et al., 2011). In fact, neither the tubular nor the flat panel photobioreactors have been demonstrated to be clearly superior to one another (Li et al., 2015).

Closed photo-bioreactors, such as column reactors, flat plate reactors, and tubular reactors, are more viable than open systems because they allow better controllability of the culture conditions, but they require in-depth knowledge of cultivation parameters and higher capital investment. Such set ups are more efficient in terms of product quality consistency that can resolve the drawbacks of open systems, but the disadvantages are overheating, cleaning issues, potential hydrodynamic stress of culture cells, and high dissolved oxygen build-up retarding the algae growth (Narala et al., 2016). Despite that, both open and closed systems deal with the same downside whereby fast-growing microalgae need complicated and expensive biomass concentration processes for commercialization of algae-based industrial products. The biomass recovery methods such as centrifugation, flocculation, flotation, sedimentation, and filtration are troublesome and costly due to the small size of algal cells (Mantzorou & Ververidis, 2019). Such challenges have motivated improvement toward traditional cultivation systems, which eventually lead to the development of state-of-the-art immobilized biofilm cultivation technology.

9.3 BIOFILM-BASED ATTACHED CULTIVATION TECHNOLOGY FOR DIATOMS

9.3.1 Dynamics of Biofilm Formation and Its Significance

A freely swimming living cell can be immobilized by natural means onto a solid substratum and is prevented from returning to the suspension phase within a biofilm. In a biofilm, cells are localized onto inanimate surfaces, and these surfaces can be in any form, including paper, stainless steel, stone, fiber bundles, cotton ducts, and plastic, as summarized in our previous work (Cheah & Chan, 2021). In nature, biofilm formation is an imperative survival strategy for the cells to adapt to a sessile lifestyle and indirectly protect themselves from the harsh environment. Microalgal biofilm is easily found on any surface with sufficient light and moisture as the embedded micro-colonies inside the 3D biofilm matrix will constantly attract other suspending cells toward their so-called "nutrient hotspot." As illustrated in Figure 9.2, biofilm development can be differentiated into five crucial stages: (1) free-floating algal cells sense a nearby substratum and come into contact with a

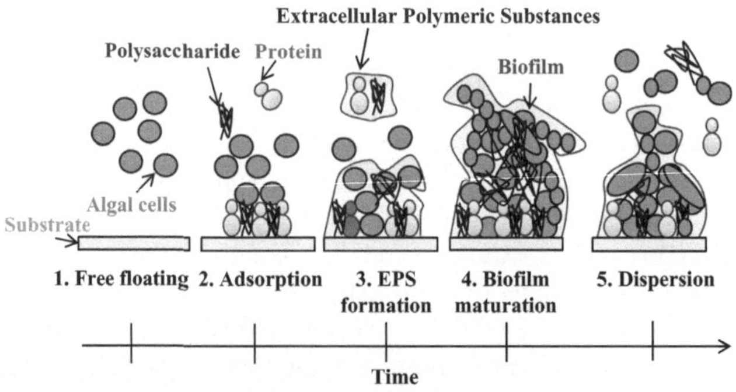

FIGURE 9.2 Microalgal biofilm development on substrates over time.

favorable substratum for cell attachment, (2) extracellular organic matters released by the suspended algal cells coat the substratum surface and create a conditioning film for the cells later on, (3) algal cells begin attachment onto the surfaces with the aid of the sticky algal organic matters, (4) cells accumulate and proliferate with adequate nutrients within the matrix, leading to increasing biofilm thickness, and (5) mature biofilm is susceptible to mass transfer limitation, hence resulting in detachment at the end of the process whereby the cells entrapped are released into the surroundings to colonize other new spots.

Diatoms are mostly the early colonizers dominating the nutrient-rich marine waters, and their attachment onto surfaces has made them key fouling organisms. They tend to adhere together by forming chain-like colonies; and on certain occasions, planktonic blooms could even be observed on the biofilm community (Patil & Anil, 2005). This biofilm formation is described as "biofouling" when it occurs in an environment where it could be nuisance. For instance, marine biofouling on ship hauls might increase the weight and frictional drag of the ship, thereby affecting its hydrodynamics, speed, fuel consumption, and maneuverability. In aquaculture, biofouling is found on either immersed infrastructure or stock organisms, such as mussels, oysters, and scallops. For marine infrastructures in oil and gas installations, heavy microalgal biofilm growth blocks the water intake into pipelines for cooling purpose, hence bringing down the efficiency of these power stations (Anil et al., 2006; Salta et al., 2013). Biofilm easily causes metal corrosion as this complex biological phenomenon can occur in a matter of minutes or hours, by the acceleration of corrosion kinetics and reactions. Extensive release of corrosion-relevant slime formers promotes tremendous growth of microorganisms – microalgae, bacteria, and fungi within biofilm communities – hence aiding the electrochemical processes and eventually leading to corrosion (Telegdi et al., 2017). Even though uncontrolled growth of biofilm seems to wreak havoc on surfaces it is attached to, with the right engineering feat, adaptation of biofilm would be a good immobilization technique for attached growth of microalgae with commercial potential.

9.3.2 Extracellular Polymeric Substances from Diatoms: Core Organic Matters within the Biofilm Matrix

Adhesion of diatoms to surfaces is inextricably linked to the exudation of extracellular polymeric substances (EPS) from the raphe slits, which lead to irreversible adhesion in the long term. EPS primarily consist of up to 90% polysaccharides, with the remaining fractions made up of proteins, lipids, and nucleic acids (Underwood et al., 2004). These macromolecules not only serve as the carbon sink for embedded cells, but most importantly, they reinforce biofilm-substrate adhesion and aid in aggregation of microalgal cells by forming a hydrated polymer network. EPS matrixes offer the host defense toward infection, aid in macromolecules digestion, and establish cell-cell communication. High molecular weight molecules of the EPS are of paramount importance in developing mucous envelopes or sheaths to create a conducive microenvironment around the diatom cells. These slimy EPS excreted through the tiny openings of the diatom keep the cell moisturized, and only the solvation of the slime will be able to liberate energy for the cells to escape from its hold. In other words, cell mobility is constrained in the presence of these sticky organic components (Shniukova & Zolotareva, 2015). In the biofilm, the presence of a slimy sulphated muco-polysaccharide, also known as transparent exopolymer particles (TEP), which is a subgroup of the algal organic matter originating from the intracellular sources of benthic diatoms, is able to promote biofilm growth in particularly under nutrient-deficient circumstances (Tong & Derek, 2021a).

These microalgal EPS are complex due to the presence of up to 10 monosaccharides and substituents, including methyl, acetyl, or sulfate groups. Their unlimited structural variability is increased because of the flexibility of the macromolecular chain, their carbon skeleton nature, their mass distribution, their molecular weight, their functional group position, their bonding type, and the rate of bonding linkage (Pierre et al., 2019). Generally, EPS can be easily distinguished into three groups based on the sites where they are found: (1) extracellular EPS, (2) structural EPS, and

(3) intracellular EPS. Extracellular EPS are sometimes termed low-molecular-weight colloidal EPS, which are weakly bounded with the algal cells and soluble in the liquid medium. The structural EPS are often known as cell-bounded EPS that closely attach around the microalgal cell surfaces in two distinct layers. These double-layered EPS are separated into loosely bounded EPS (LB-EPS) and tightly bounded EPS (TB-EPS). The former are normally entangled by weak interactions, whereas TB-EPS are responsible for cell adhesion in the inner biofilm matrix via strong interactions. Intracellular EPS are referred to as internal organic matter of microalgae due to cell lysis or cell wall disruption. By employing appropriate extraction procedures, different types of EPS can be easily isolated by physical, chemical, and a combination of both methods (Cheah & Chan, 2022). This alteration in the EPS composition is dynamic according to the microalgae species, culture conditions, and cell age. For instance, the chemical derivation of the EPS of *Fragilariopsis cylindrus* was more diversified at lower temperatures with increased proportions of mannose, fucose, and galactose in dissolved EPS (Aslam et al., 2018). *Amphora* sp. was depicted to produce a higher fold of total exopolysaccharides, but a lesser amount of tightly bounded EPS than *Stauroneis* sp. (Jin et al., 2018). Both the EPS polysaccharide and protein concentration increased with temperature for three marine benthic diatoms (*Amphora coffeaeformis*, *Cylindrotheca fusiformis*, and *Navicula incerta*), but the opposite trend was observed for internal organic matter (Cheah et al., 2021). It is also postulated that the EPS level will most likely be upregulated under extreme conditions since they are excreted as a defense mechanism against such threats.

9.4 FUNDAMENTAL DESIGNS OF BIOFILM PHOTOBIOREACTORS FOR DIATOMS' MASS CULTURE

A biofilm photobioreactor is a vessel used to grow biofilm and to house non-porous or porous surfaces as supporting materials for the microalgal biofilm. It is designed to mimic the natural biofilm environment, and different cultivation conditions can be imposed to measure the response of biofilm based on the respective research interests. Surfaces known to be used to grow biofilm include hard non-porous surfaces such as metals, glass, and plastics, or porous surfaces likes cement, tile, textiles, and porcelain (Goeres et al., 2020). As summarized by Goeres and others (2020), in laboratory studies over the years, biofilm photobioreactors can be in varying configurations such as 96-well plate reactor, drip flow biofilm reactor, rotating disk biofilm reactor, bio-inline biofilm reactor, microfluidic devices, flow lane incubators, and biofilm annular reactor. Reactor choices depend highly on the targeted applications. They have been extensively utilized and suited to specific applications such as, to determine the optimum substrates for biofilm culture (flow lane incubators and drip flow biofilm reactors), to examine the chemical resistance of the biofilm culture (96-well plate biofilm reactors), or for large-scale environmental applications (rotating disk biofilm reactor). These biofilm photoreactors can be operated under fully moistened, partially hydrated, or dry mode by manipulating the arrangement of the solid substrates to mimic the fluid dynamics in the environment of interest.

Generally, photobioreactor designs are classified into three groups: constantly submerged biofilm systems, partially submerged biofilm systems, and permeated biofilm systems. A schematic diagram of a fully submerged biofilm reactor system is displayed in Figure 9.3(a). The selected substrates are permanently submerged in the liquid medium to grow the microalgal biofilm. An example of such an application is shown in Figure 9.3(b) using commercial polyethersulfone membranes with two different surface roughness as the main substrates to grow the benthic biofilm for 20 days under a dynamic flow condition. The attached microalgal culture can also be performed in multi-layer submerged photobioreactors (Shen et al., 2015). In another study, marine diatom biofilm was cultured on polycarbonate slides in a flow lane incubator proliferated until reaching full substratum coverage at high irradiances and continuous liquid flow (Guzzon et al., 2021). Diatoms readily form biofilm onto any exposed substrates in a water column within a short time span, as demonstrated in a petri-dish biofilm reactor, whereby three different marine diatoms colonized the commercial membranes rapidly within hours due to the secretion of EPS (Tong &

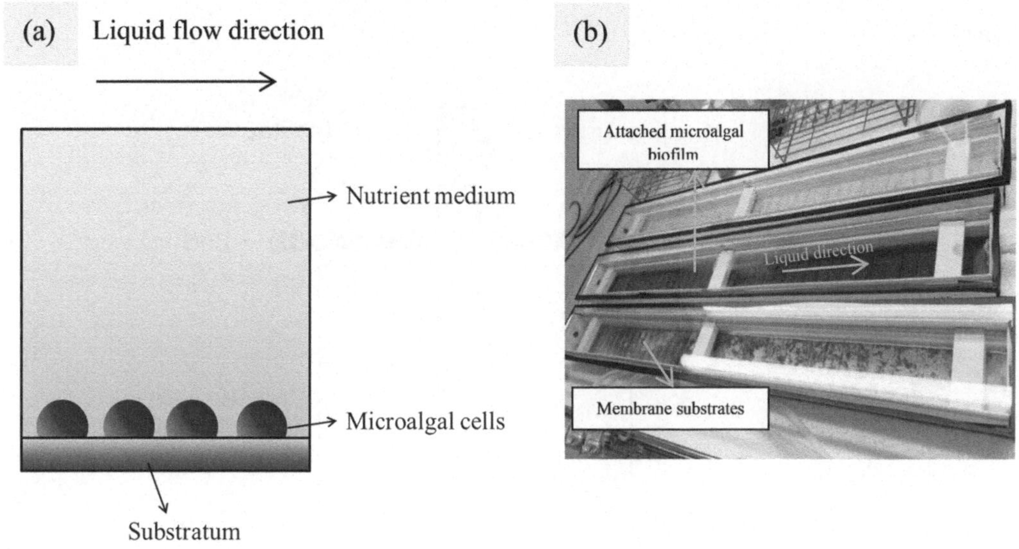

FIGURE 9.3 Constantly submerged biofilm systems: (a) Schematic diagram, and (b) Flow lane incubator that was used to culture three different marine diatom species: *Amphora coffeaeformis*, *Cylindrotheca fusiformis*, and *Navicula incerta* (Tong & Derek, 2022).

Derek, 2021a). However, submerged biofilm cultivation systems are often affected by strong shear liquid force, particularly when the biofilm matured and begin to lose its structural stability. Therefore, the influence of substratum surface properties is of great importance as it was observed that diatoms, together with other microalgae species, form a stronger biofilm on rougher and hydrophobic surfaces (Al-Amshawee et al., 2020; Ozkan & Berberoglu, 2013; Tong et al., 2021; Tong & Derek, 2022; Tong & Derek, 2021b; Zhang et al., 2020).

The supporting materials can be partially submerged in a tank with a motor-driven rotating axis to intermittently expose the biofilm attached between ambient air and aqueous nutrient broth, as shown in Figure 9.4(a). When the biofilm is out of the water, the exposure to the carbon dioxide in the air promotes the biofilm growth. In a rotary disk reactor, several substrate discs can be hold together in a row and attached to a support with adjustable rotational speed (Azeredo et al., 2017). The rotating discs are typically submerged 40% below the water level and operated at a slower speed to avoid biofilm exfoliation under vigorous mechanical movement. These rotating supports are usually circular shape to assure a large surface area in a relatively narrow space, but sufficiently separated from each other to prevent liquid channel obstruction due to extensive biofilm formation (Mizyed, 2021). A variable speed motor is used to optimize the rotation frequency of the axle to acquire a well-mixed liquid phase and consistent hydrodynamic conditions (shear stress, etc.). This rotating biological contactor turned out to be a viable system in secondary wastewater treatment, as reviewed by a recent study in which healthy photosynthetic biofilm with high phycocyanin yield was cultured on oilfield and natural gas–produced wastewater (Wood et al., 2022). Such a system also promotes the growth of other beneficial organisms, as seen in the study by Zhou et al. (2018) for acid mine drainage water treatment. The dominant diatom species of *Nannochloropsis* sp. and *Nitschia* sp. at day 1, was diversified with various acidophilic bacteria and eukaryotic algae co-existing in the biofilm as the treatment progressed (Zhou et al., 2018).

As illustrated in Figure 9.4(b), another possible design that shares a similar concept with partially submerged biofilm reactors is called a rocking cultivation reactor (Shen et al., 2014). In this rocking system, a transmission arm is attached at the center to provide a small tilting

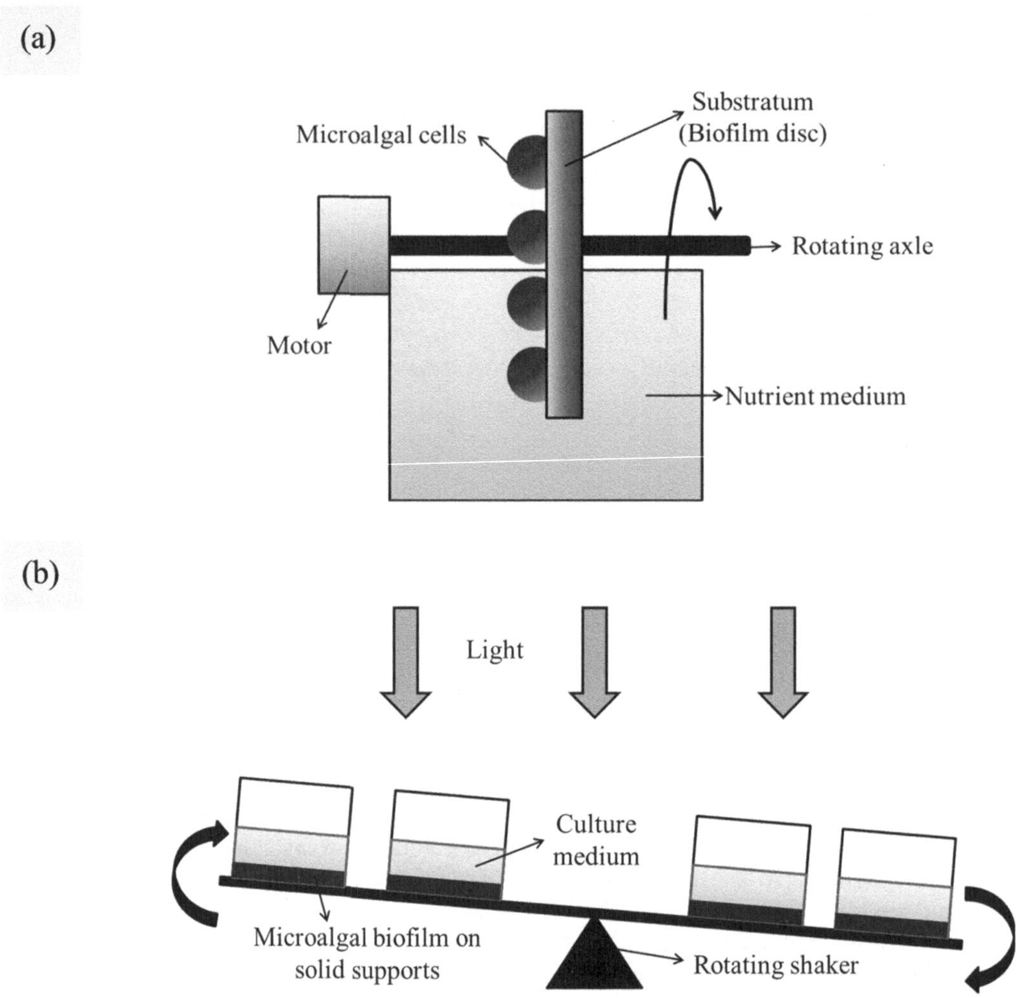

FIGURE 9.4 Partially submerged biofilm systems: (a) Schematic diagram, and (b) Rocking biofilm cultivation photobioreactor.

movement. The tilting movement is greater when the arm was attached to the hole further from the center. Gas exchange occurs when the biofilm is exposed to the atmosphere during the rocking motion of the photobioreactor. The required reciprocal motion of the algal culture can be generated by integrating the motion with the rocking frequency (Skjånes et al., 2016). Generally, these partially submerged biofilm systems offer short contact periods between the microalgae and the liquid medium and are capable of handling a wide range of inlet flow. Nonetheless, they are often challenged by high difficulty in infrastructure maintenance and complicated construction. Process flexibility and scaling-up ability are sometimes limited, and there may be slow process start-up.

Efforts to further enhance the biofilm attachment is achieved by fully excluding the nutrient broth from the biofilm surface and putting it in direct contact with ambient air. Such a biofilm system emphasizes a core component, a thin and porous sheet-like material to support the microalgae (Podola et al., 2017). As depicted in Figure 9.5(a), it relocates the bulk culture medium to the posterior of the biofilm while facilitating nutrient diffusion via the micropores of the porous supports. There are two operational modes in a porous substrate biofilm reactors: (1) bulk liquid is circulated behind the porous sheet (substrate layer), and (2) liquid is uniformly distributed through

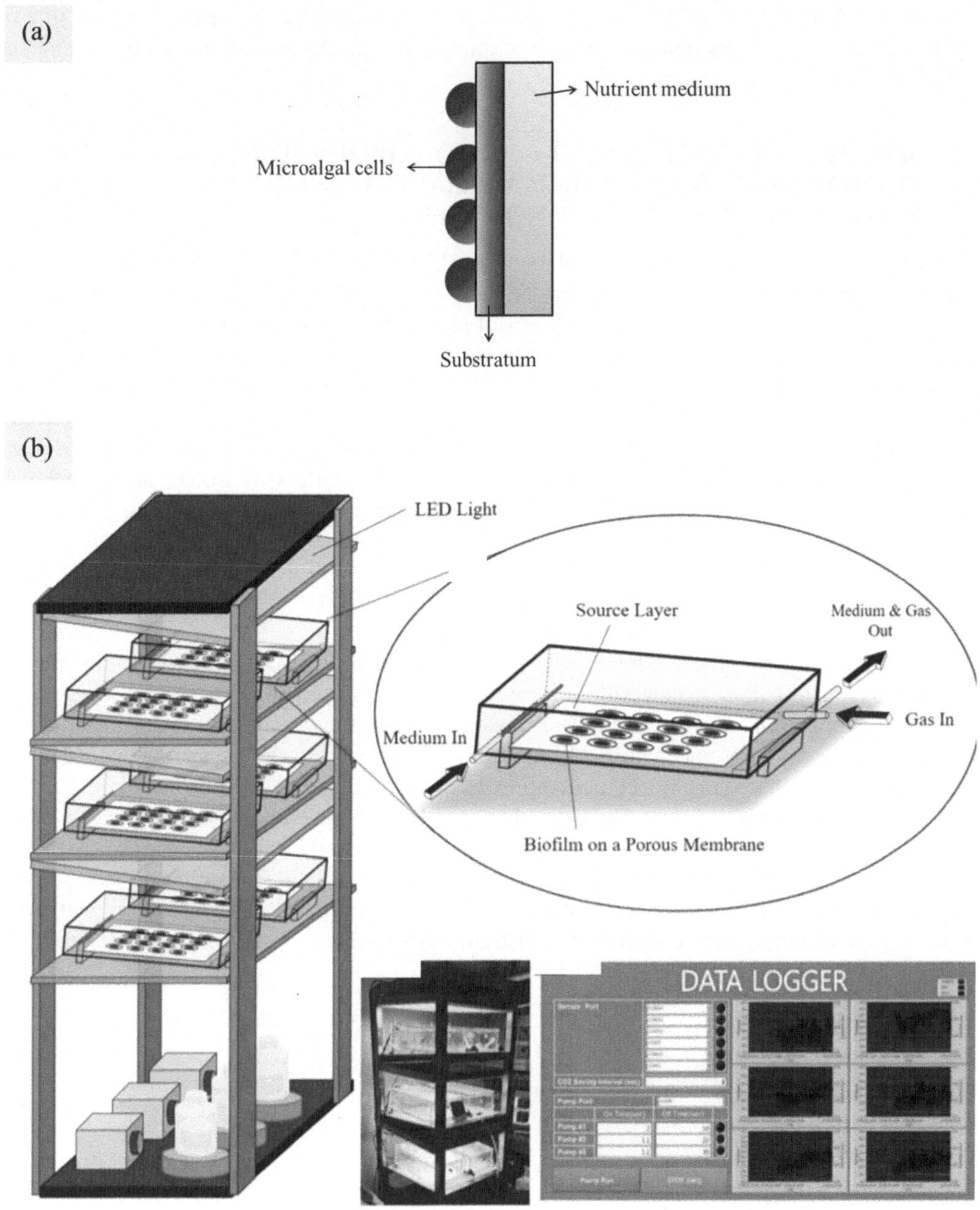

FIGURE 9.5 Permeated biofilm systems: (a) Schematic diagram, and (b) Detailed image and actual photo of an entire porous substrate photobioreactor coupling with a real-time data logging program of temperature, humidity, and carbon dioxide concentration inside of each bioreactor (Kim et al., 2018) (CC BY).

gravitational flow via means of a second macroporous or nonwoven sheet-like structure, such as paper or filter cloth (source layer), whereby this cultivation method is sometimes termed the "twin-layer system" (Figure 9.5(b)). A maximal microalgal biomass value of about 232 g m^{-2} was attained after 33 days of biofilm growth in a twin-layer system (Carbone et al., 2020). Similarly, by merely using a small microalgal inoculum size (2.5 g m^{-2}), a maximum biomass surface productivity of approximately 28 g m^{-2} day^{-1} was achieved under optimal culture conditions

(Kim et al., 2018). Compared with submerged or partially submerged systems, this principle of separation not only allows denser biofilm accumulation, but also minimizes both detachment and contamination risk.

9.5 SPECIAL HIGHLIGHT ON BIO-COATED POROUS SUBSTRATE BIOFILM PHOTOBIOREACTORS (BPSBR): A PARADIGM SHIFT IN MICROALGAL BIOFILM CULTIVATION TECHNOLOGIES

Bio-coated porous substrate biofilm photobioreactor (BPSBR) is described as a novel biofilm cultivation chamber with the aid of bio-coating derived from the microalgae themselves. A bio-coating is a pre-deposited coating comprising thin layers of microalgal organic matters, EPS to enhance the cell immobilization in the biofilm. Inspired by the conditioning film formation whereby the EPS will adsorb onto the solid surfaces during biofilm development, this pre-deposited EPS layer is able to serve as a nutritional platform to effectively anchor the planktonic cells. It is non-toxic, chemically stable, environmentally friendly, inexpensive, and most importantly, EPS bio-coating has high compatibility with the algal cells, resulting in a high cell viability rate. As illustrated in Figure 9.6, in BPSBR, selected solid substrates are immersed or coated in extracted cell-free EPS solution overnight to establish a thin bio-coating onto the surfaces prior to microalgal biofilm cultivation to any type of biofilm photobioreactor settings, as discussed in **Section 9.4**. This bio-coating can be easily applied in fully submerged, partially submerged, or permeated systems without any complicated preparation steps and regardless of the substrates' type.

The BPSBR design offers a huge difference with all the well-established biofilm photobioreactors as it provides a new solution toward the biofilm sloughing issue that frequently occurs in natural biofilm cultivation. Biofilm that develops naturally on clean substrates is susceptible to exfoliation under nutrient deprivation, especially when the biofilm is grown on unfavorable substrates. This novel cultivation technique not only short-circuits the need of natural biofilm development, but also strengthens the biofilm adherence with uniform coverage over the substrates. Moreover, this approach has also shown good performance in growing EPS-lacking microalgae using EPS-rich strains (Nongmaithem et al., 2021). Having a stark difference from the other earlier-described biofilm photobioreactors, the BPSBR cultivation method is an alternative by turning any unfavorable substrates into nutritional platforms via EPS masking,

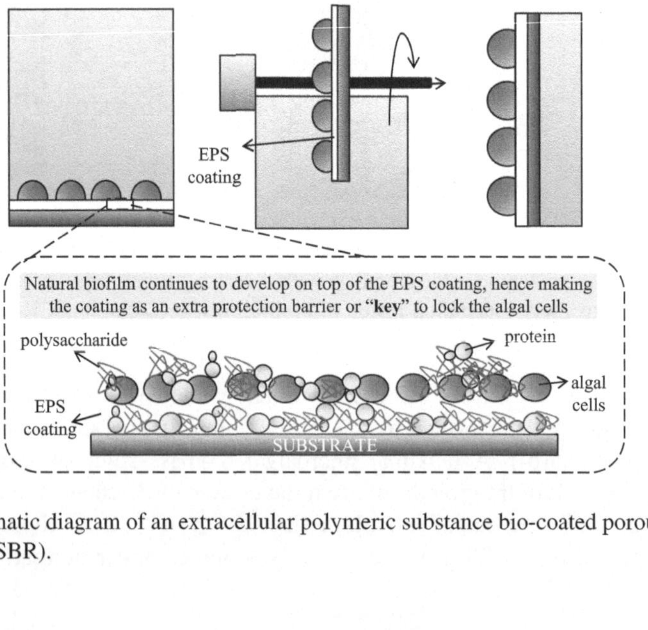

FIGURE 9.6 Schematic diagram of an extracellular polymeric substance bio-coated porous substrate biofilm photobioreactor (BPSBR).

as well as catering to the needs of cultured microalgae via optimization of the bio-coating formulation (EPS volume, EPS concentration, and pH etc.). It appears to be an effective biofilm-based platform that not only outperforms both suspension control systems, but also traditional biofilm-based systems.

Being a well-known EPS producer, diatom is a good candidate in BPSBR as their self-produced EPS have a porous microstructure to easily entrap other living algal cells and impart osmo-protectant properties to the cells (Tong & Derek, 2021a). However, research is still scarce with regards to diatom biofilm cultivation in the presence of bio-coating. Studies relevant to this attractive feature are currently underway to lead a more nuanced view of cell immobilization. For example, EPS pre-coated polyvinylidene fluoride membrane substantially improved the cell attachment rate of the marine microalgae, *Navicula incerta*, to at least 10 times greater than the initial cell adhesion degree on pristine surfaces (Tong et al., 2022). A rough substrate surface pre-coated with EPS biofilm derived from 12-day-old *Limnothrix* sp. has shown to promote a maximum growth for *Scenedesmus* sp. with aerial biomass productivity of 31.6±1.20 g m^{-2} day^{-1} (Nongmaithem et al., 2021). There was also a study investigating the role of bacterial EPS on enhancement of attached microalgae growth in a biofilm-based photobioreactor. Findings have shown a significant improvement in biomass attachment of microalgae *Scenedesmus*. LX1, as high as 230% compared with uncoated substrates, with each single attached cell having higher protein content at the late stages of the batch cultivation (Zhuang et al., 2016).

9.6 KEY SUBSTRATE PROPERTIES AFFECTING BIOFILM GROWTH AND ADHESION IN BIOFILM PHOTOBIOREACTORS

In any biofilm photobioreactors, colonizing microalgae sense and respond to different surfaces in terms of their respective settlement dynamics. Given the broad spectrum of environmental factors affecting the biofilm formation process, surface characteristics of the supporting materials, such as surface chemistry, surface hydrophobicity, surface roughness, surface topography, surface charges, and surface coating, each have their own role in microalgae adhesion, as illustrated in Figure 9.7. Myriad features on the substrate not only offer different contact surface areas, but also induce subtle variations of physico-chemical interactions between the microalgae and the pristine substratum.

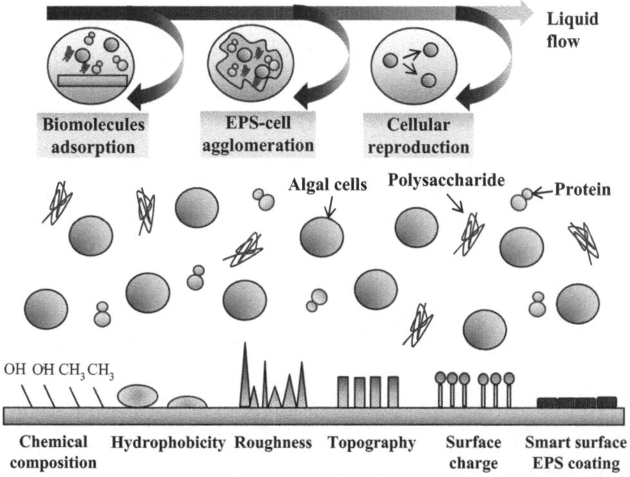

FIGURE 9.7 Implication of surface properties on cell adhesion and biofilm formation (modified from (Song et al., 2015)).

9.6.1 Surface Chemistry and Charge

Surface chemistry is a type of chemical phenomena occurring at the interface of two surfaces, which can be any combination of phases (i.e., solid-liquid, solid-gas, liquid-gas), and it is often associated with molecule adsorption in many biological systems. It affects the biofilm development through surface charge modifications by immobilization of conductive or electroactive polymers, altering the amount of electric charges accumulated over a surface. Simultaneously, physical (thermal) and chemical treatments can alter the biofilm formation rate by changing the surface morphology and hydrophilic moieties. For instance, silicon surfaces modified with hydrogen peroxide and hydrochloric acid created a hydrophilic (hydroxyl group-terminated), and this resulted in less cell attachment of *Psuedomonas aeruginosa* (Zhang et al., 2018a). Functional groups can be also grafted to a solid surface to modify the surface's chemical properties, affecting the degree of cell anchorage.

Generally, microalgae are always encapsulated in a shell of sulfated polysaccharides, proteins, nucleic acids, and phospholipids to shield themselves from environmental stress, hence giving the cells a negatively charged surface (Zhao et al., 2021). Therefore, it was hypothesized that cell adhesion can be governed by substrate surface charge-switch, pH-responsive modifications, or functionalization to generate electrostatic attraction or repulsion. A negatively charged surface depicted a reduction in cell settlement as compared to positively charged polytetrafluoroethylene substrates (Rosenhahn et al., 2009). Similarly, polyethylenimine (PEI), a biocompatible polymer, is known to be positively charged at low pH (<7), which facilitates cell adhesion. As an example of its application, the anode of microbial fuel cell is often treated with PEI to enhance the fuel cell performance, resulting in higher power density (Karthikeyan et al., 2016). A PEI-cross-linked patterned PVDF membrane was also proven to be positively charged at pH < 8 and negatively charged at pH > 8. Such variation in the substrate surface charge at different pHs lead to optimal biofilm formation at pH 7 and eased microalgal harvesting at pH above 9 (Zhao et al., 2021).

9.6.2 Surface Hydrophobicity

Surface hydrophobicity is a physical measurement of the surface's affinity toward water. It is generally understood that the interfacial tension of water with a solid (algal cells or solid substrates) has a significant impact on the interaction of the submerged surface with biomolecules and colloids from EPS. A substrate having lower surface tension (higher water contact angle) favors cell attachment; however, it was found that the cell adhesion does not linearly increase with the surface energy of the substrate, but there is a certain surface energy range where the cell adhesion is minimal. When the substrate's surface contact angle is larger than 60°, an increase in cell adhesion is observed (Rosenhahn et al., 2010). Ozkan et al. (2013) revealed that greater adhesion strength was expected for hydrophobic microalgal cells on hydrophobic indium tin oxide–coated substrata using classical colloidal approaches (Ozkan & Berberoglu, 2013). Nevertheless, caution needs to be taken as the organic matters or so-called bio-surfactants (from EPS) present in the liquid solution might render variation of the surface tension of the liquid itself. With decreasing surface tension of the liquid medium, the microalgal biofilm development on a hydrophilic surface was slow initially but picked up once hydrophobic EPS started to coat the surface (Zhang et al., 2018b). As cited in an earlier study, a lack of EPS production caused low adhesion of the cells toward substrates, conferring a semi-planktonic state on the microorganism under biofilm-forming conditions (Forson et al., 2020).

It can be generalized that irreversible adhesion tends to occur via the hydrophobic interactions due to the water exclusion mechanism between the cells and solid substrates. Hydrophilic surfaces are thought to facilitate protein desorption because proteins are shielded from the surface by a high-density water layer (Firkowska et al., 2017). If the kinetic energy of the microalgal cells is greater than the overall contact energy between the solid surfaces and the cells, the initial adhesion may be reversible (Soriano-Jerez et al., 2021). It is common to observe that a hydrophobic surface gives irreversible adhesion at a great magnitude of primary energy minimum over a short range of small separation distances, fostering an energetically favorable condition for the algal cells to attach on the hydrophobic supporting materials. Nevertheless, the effect of substrate surface hydrophobicity toward microalgal biofilm formation has so far remained divisive as no significant trends were observed in the study between hydrophilic glass and hydrophobic polytetrafluoroethylene for the microalgal colonization of *Scenedesmus obliquus* (Edward, 2010). A similar conclusion was also drawn proposing that simply because a solid surface is hydrophilic does not indicate it is protein resistant due to the synergistic effect of the surface hydrophobicity and surface charge (Soriano-Jerez et al., 2021).

9.6.3 Surface Roughness

It is well known that surface roughness level plays an essential role in cell adhesion and biofilm formation, but there is no "one-size-fits-all" rule between surface roughness and cell attachment due to the huge range of variation in the cell sizes and shapes (Song, 2016). The most straightforward method for predicting algal adhesion to surfaces is to screen a wide range of materials for their effects on the adhesion. A study was done by Zhang et al. (2020), who selected pine sawdust, sugar bagasse, rice husk, elastic filler, and quartz sand as the main supporting materials in a long-term microalgal biofilm cultivation (Zhang et al., 2020). Their study revealed that, strengthened and robust microalgal biofilm was found on rougher surfaces due to the presence of more asperities and hydrodynamic stagnant zones that are able to promote cell retention. A higher amount of *Nitzschia amphibia* cells were found to attach onto rougher substrate surfaces (Sekar et al., 2004). Interestingly, biofilm structure can be varying on solid surfaces with different roughness, but a biofilm with high coverage ratio and greater uniformity was presented on a smoother surface (Champigneux et al., 2019). On a rougher structure, the algal biofilms appeared to be denser and thicker as the "valleys" found on the rough surfaces shelter the suspended microalgae, and this leads to an irreversible contact adhesion that can be further furnished with EPS release (Tang et al., 2021). In such a context, microalgae colonies rapidly form and multiply, ending up with higher biomass accumulation.

9.6.4 Surface Topography

The topography of a microhabitat is defined as the irregularities on the surfaces of the supporting materials due to depressions, crevices, and protrusions (Souza & Ferragut, 2012). Recent advancements in material and surface engineering have opened up new possibilities for creating surfaces with well-defined topography patterns. Grooves, ridges, pillars, hills, and pores with defined dimensions are examples of functionally designed surface configurations. *Nannochloropsis oculata* and *Scenedesmus dimorphus* attached the most to both nylon and polycarbonate micro-engineered with groove-shaped surface texture due to higher contact points (Cui et al., 2013). Besides, the V-grooved surface (0.00105 to 0.01092 Pa) depicted a lower shear stress than that on the U-grooved surface (0.00108 to 0.03609 Pa), and the initial *Scenedesmus obliquus* cell attachment time was reduced to 50 minutes as compared to 135 minutes on unmodified smooth surfaces (Huang et al., 2018). These inerratic microgrooves were able to function as anchor points, evoking larger contact areas for microalgal adhesion, and created vortex-free zones to shelter the cells from fluid shear stress. A streamline distribution is always preferred as improper engineered groove angles can cause the liquid medium to flow unhindered into the grooves, and the outflowed can easily carry away the detached cells from the surfaces, as presented in 60° V-grooves for Scenedesmus obliquus (Huang et al., 2018).

These studies also suggested that functional-designed topography enhanced microalgal cells' attachment on specific size dimples according to the cell sizes and shapes. As an example,

Nannochloropsis oculata, a sphere-shaped cell with 2–5 μm diameter that was close to the designed ridge spacing size of about 5 μm width, was expected to have higher cell colonization degree, but the attachment process was limited by the ridge depth with only 1 μm, giving fewer potential contact points (Cui et al., 2013); 500 μm sections were shown to be the best featured size for benthic microalgae colonization in such inert zones introduced by the adjacent 3D-printed hemispheres (Kardel, 2016). The flow turbulence characteristics within the fluid boundary layer are determined by the heterogeneity of the substratum morphology, which influence the kinetics of cell colonization, growth, and metabolism (Kardel, 2016). Different shape morphologies and sizes of the substratum change the velocity gradient at boundary layers, establishing varying flow patterns at the surfaces. Three key distinct boundary regions can be defined based on the cell-substratum separating distances: (1) a fluid boundary layer, which is slightly turbulent to the biofilm surface and more susceptible to biofilm exfoliation (>20 μm), (2) a secondary minimum area with fewer hydrodynamic effects that easily leads to reversible cell adhesion (10–20 μm), and (3) a free-energy barrier layer, where irreversible cell adhesion happens (<10 μm). Micro-topography on the substrates allows the bulk nutrient flow to reach inter-cellular regions toward the biofilm-attached substratum, whereas the nutrient and oxygen flux can be facilitated by the pore-sipping mechanism within the heterogeneous 3D biofilm matrix or the substrates surface itself. Therefore, a significant impact on the biofilm growth kinetics could be influenced by the boundary layer flow characteristics that affect advection and diffusion of limiting nutrients (Krsmanovic et al., 2021).

9.7 BENEFITS AND APPLICATIONS OF BIOFILM PHOTOBIOREACTOR SYSTEMS

Commercial applications of the diatom biofilm have been extensively reviewed over the years, as projected in Table 9.1. Diatoms are widespread in the majority of hydrosystems and have high toxin tolerance. They are good candidates for wastewater treatment in the form of biofilm since they uptake nitrate, orthophosphate, silica, and an array of heavy metals for growth. Owing to their diversity and trophic flexibility, diatoms gain an advantage over other microalgae when cultivated as biofilm in wastewater treatment (Chaib et al., 2021). Complex composition of the wastewater might sometimes limit the biomass yield as not all microalgae are able to adapt themselves in excessively contaminated waters, in particularly with hazardous chemicals; but diatoms with a strong survival rate in a wide range of salinity are worthy of consideration. Wastewater has to be diluted to bring down the pollutant level to allow for microalgae growth; however, dilution with clean fresh water was unsustainable due to its scarcity. In this context, seawater was utilized instead as a cost-cutting measure, which worked out well for cultured diatoms (Mishra & Mohanty, 2019). Other than that, biomass obtained from wastewater remediation can be further used for the production of biofuels, nutraceuticals, biomolecules, pigments, feed additives, and bio-fertilizers. For instance, *Phaeodactylum tricornutum* and *Navicula pelliculosa* were able to remove dichromate while concomitantly producing lipids (Hedayatkhah et al., 2018).

Diatoms can produce lipids in quantities comparable to or larger than cyanobacteria and filamentous green algae. Under normal growth conditions, diatoms will have a lipid content of about 22.7% of dry cell weight and rise to 44.6% under stress (Marella et al., 2020). They accumulate a notable amount of long-chain fatty acids in the form of omega-3-linolenic acid, eicosapentaenoic acid, docosahexaenoic acid, arachidonic acid, and linoleic acid, which is always associated with health benefits for human beings, especially in reducing the risk of cardiovascular disease and improving brain development (Matos, 2017). Marine diatom, *Thalassiosira weissflogii*, was mass cultured in outdoor bioreactors via the combined effects of physiochemical conditions for eicosapentaenoic acid and fucoxanthin production (Marella & Tiwari, 2020). In aquaculture, diatoms serve as high-quality feed for larvae and shrimp; the high vitamin and amino acid content benefits aquaculture. Macroalgae powder was used as a suspending carrier to culture the benthic diatom, *Nitzschia* sp., followed by feeding of juvenile *Stichopus japonicas* with the cultured diatoms (Jiang et al., 2020). However, a wide range of secondary metabolites with antifungal, antioxidant, and

TABLE 9.1
Diatoms-based Biofilm Applications

Diatoms	Applications	Type of Photobioreactor	Remark	Ref.
Diatom consortium	Wastewater bioremediation	Batch-mode fixed bed bioreactor filled with different sizes of coarse sand	The maximum removal rates were quantified as 75.26% for ammonium, 72.43% for nitrate, and 61.65% for phosphorus.	(Chaib et al., 2021)
Nitzschia sp.	Aquaculture feed	Macroalgae powder	The growth rate of sea cucumbers fed with macroalgae-cultured diatoms was about 1.5 times higher than that of ones fed with commercial feed.	(Jiang et al., 2020)
Pennate diatoms	Urban wastewater treatment	Submerged biofilm photobioreactor with polyester mat as a main substratum	High nitrogen (2.52 g m^{-2} d^{-1}) and phosphorus (1.25 g m^{-2} d^{-1}) removal rates were obtained with a marked increase in the volume of lipid bodies (9.29 g m^{-2} d^{-1}) inside diatom cells.	(Marella et al., 2019)
On-site diatom biofilm	River water quality assessment with lipid production within the diatom bodies	Benthic diatom biofilm scraped directly from concrete walls in the selected industrial effluent sites	A high number of lipid bodies was discovered in the diatoms found from contaminated sites, although there were valve deformities and a higher proportion of dead frustules.	(Pandey et al., 2018)
Nitzschia thermallis-dominated microbial biofilm	Phototrophic carbon capture system	Flat plate photobioreactor	Under blue light, biofilms were dominated by diatoms and displayed robust biomass productivity about 80 days. Biofilm cultured at high pH and alkalinity was shown to be a stable phototrophic carbon capture system.	(Sharp et al., 2017)
Nitzschia palea	Lipid accumulation	Flat plate parallel horizontal photobioreactor with glass substrates inside	Diatoms formed more robust biofilms with higher biomass and lipid productivities than green microalgae.	(Schnurr et al., 2013)
Navicula veneta and bacteria	Mass cultures of marine invertebrates for commercial purposes	Submerged biofilm reactor with sterilized polyethylene bristles in a "bottle brush" arrangement	Navicula veneta biofilm production was optimized to improve the settling of marine larval shellfish by incorporating bacteria into the culture, resulting in a cell production of >50% of the control group.	(Avendaño-Herrera & Riquelme, 2007)

antibiotic properties from diatoms such as alkanes, terpenes, and sterols are not fully explored yet (Marella et al., 2020). To date, applications of diatom-based biofilms are still under-reported, especially investigations regarding the effects of additives or chemicals toward diatoms' growth and biochemical profile.

9.8 CONCLUSION

Diatoms, nature's silicon marvel with their exquisite structural designs and high cell adaptability, have underscored the importance of more detailed scientific investigations, particularly in terms of biofilms. The extracellular polymeric substances secreted by the diatoms have shown a superior tendency to adsorb onto any surfaces and aid the development of biofilm photobioreactors. Despite all the environmental attributes described earlier, diatom biofilms are still poorly understood in terms of their interactions among the cells or supporting substrates. In view of this eco-friendly and sustainable option, more innovative means should be established to make diatoms the feedstock of a plethora of high-value products, ranging from biofuels, bioremediation, bioactive compounds, nanotechnology, biomineralization, bioelectricity generation, and drug delivery systems. Owing to the benefits that surpassed conventional suspension-based cultivation systems, huge-scale biofilm-based photobioreactors are a promising and sustainable approach to mass culture diatom biofilms for commercial products in the future.

ACKNOWLEDGEMENT

Acknowledgement to Universiti Sains Malaysia for Research University Grant with Project Code: RUI/8014065.

REFERENCES

Al-Amshawee, S., Yunus, M. Y. B. M., Lynam, J. G., Lee, W. H., Dai, F., Dakhil, I. H., 2020. Roughness and wettability of biofilm carriers: A systematic review. *Environ Technol Innov* 21, 101233.

Anil, A., Patil, J., Mitbavkar, S., D'Costa, P., D'Silva, M. S., Hegde, S., Naik, R. K., 2006. Role of diatoms in marine biofouling. *Recent Advances on Applied Aspects of Indian Marine Algae with Reference to Global Scenario* 1, 351–365. https://drs.nio.org/drs/bitstream/handle/2264/810/Recent_Adv_Appl_Aspects_Indian_Mar_Algae_2006_1_351.pdf?sequence=2

Aslam, S. N., Strauss, J., Thomas, D. N., Mock, T., Underwood, G. J. C., 2018. Identifying metabolic pathways for production of extracellular polymeric substances by the diatom *Fragilariopsis cylindrus* inhabiting sea ice. *ISME J* 12, 1237–1251.

Avendaño-Herrera, R. E., Riquelme, C. E., 2007. Production of a diatom-bacteria biofilm in a photobioreactor for aquaculture applications. *Aquacultural Eng* 36, 97–104.

Azeredo, J., Azevedo, N. F., Briandet, R., Cerca, N., Coenye, T., Costa, A. R., Desvaux, M., Di Bonaventura, G., Hebraud, M., Jaglic, Z., Kacaniova, M., Knochel, S., Lourenco, A., Mergulhao, F., Meyer, R. L., Nychas, G., Simoes, M., Tresse, O., Sternberg, C., 2017. Critical review on biofilm methods. *Crit. Rev Microbiol* 43, 313–351.

Burgess, S. J., Tamburic, B., Zemichael, F., Hellgardt, K., Nixon, P. J., 2011. Chapter 3 - Solar-driven hydrogen production in green algae. In: Laskin, A. I., Sariaslani, S., Gadd, G. M. (Eds.), *Adv. Appl. Microbiol.* Academic Press, pp. 71–110.

Cameron, N. G., 2013. Diatoms. In: Elias, S. A., Mock, C. J. (Eds.), *Encyclopedia of quaternary science.* Elsevier, Amsterdam, pp. 522–525.

Carbone, D. A., Olivieri, G., Pollio, A., Melkonian, M., 2020. Biomass and phycobiliprotein production of *Galdieria sulphuraria*, immobilized on a twin-layer porous substrate photobioreactor. *Appl Microbiol Biotechnol* 104, 3109–3119.

Casini, D., Prussi, M., Chiaramonti, D., Rodolfi, L., Bassi, N., Bacci, F., Bondioli, P., Della Bella, L., Rivolta, G., Tredici, M. R., 2011. Improved raceway ponds for microalgae production. In: ISAF - XIX International Symposium on Alcohol Fuels "Development and utilisation of alcohol fuels, to promote sustainability". Verona.

Chaib, N., Dzizi, S., Kaddeche, H., Noune, F., 2021. Performance of a fixed-bed bioreactor using diatom biofilms for wastewater bioremediation. *CLEAN – Soil, Air, Water* 49, 2000282.

Champigneux, P., Renault-Sentenac, C., Bourrier, D., Rossi, C., Delia, M.-L., Bergel, A., 2019. Effect of surface roughness, porosity and roughened micro-pillar structures on the early formation of microbial anodes. *Bioelectrochemistry* 128, 17–29.

Cheah, Y. T., Chan, D. J. C., 2022. A methodological review on the characterization of microalgal biofilm and its extracellular polymeric substances. *J Appl Microbiol* 132, 3490–3514.

Cheah, Y. T., Chan, D. J. C., 2021. Physiology of microalgal biofilm: A review on prediction of adhesion on substrates. *Bioengineered* 12, 7577–7599.

Cheah, Y. T., Lakbir Singh, H. K. K., Chan, D. J. C., 2021. Effect of high temperature toward microalgal organic matter and its impact toward membrane distillation application. *Water Environ Res* 93, 1107–1115.

Cheah, Y. T., Ng, B. W., Tan, T. L., Chia, Z. S., Chan, D. J. C., 2022. Biomass and eicosapentaenoic acid production from Amphora sp. under different environmental and nutritional conditions. *Biotechnol Appl Biochem* 70, 568–580. https://pubmed.ncbi.nlm.nih.gov/35767864/

Cui, Y., Yuan, W., Cao, J., 2013. Effect of surface texturing on microalgal cell attachment to solid carriers. *Int J Agric Biol Eng* 6, 44–54.

Djamal, Z., Henni, A., 2019. Outdoor microalgae cultivation for wastewater treatment, *Application of Microalgae in Wastewater Treatment*. Springer, New York, pp. 81–99.

Edward, I. T. 2010. Factors influencing the formation and development of microalgal biofilms., *Vol. Master*. University of Toronto.

Firkowska, I., Zhang, X., Jandt, K., 2017. Controlling protein adsorption through nanostructured polymeric surfaces. *Adv Healthcare Mater* 7, 1700995.

Forson, A. M., van der Mei, H. C., Sjollema, J., 2020. Impact of solid surface hydrophobicity and micrococcal nuclease production on *Staphylococcus aureus* Newman biofilms. *Sci Rep* 10, 12093.

Goeres, D. M., Pedersen, S., Warwood, B., Walker, D. K., Parker, A. E., Mettler, M., Sturman, P., 2020. Chapter 3 - Design and fabrication of biofilm reactors. In: Simoes, M., Borges, A., Chaves Simoes, L. (Eds.), *Recent Trends in Biofilm Science and Technology*. Academic Press, pp. 71–88.

Guzzon, A., Di Pippo, F., Bonavita, S., Congestri, R., 2021. Influence of light and flow on taxon composition and photosynthesis of marine phototrophic biofilm in photobioreactors. *Mar Environ Res* 169, 105395.

Hedayatkhah, A., Cretoiu, M. S., Emtiazi, G., Stal, L. J., Bolhuis, H., 2018. Bioremediation of chromium contaminated water by diatoms with concomitant lipid accumulation for biofuel production. *J Environ Manag* 227, 313–320.

Hildebrand, M., Davis, A. K., Smith, S. R., Traller, J. C., Abbriano, R., 2012. The place of diatoms in the biofuels industry. *Biofuels* 3, 221–240.

Hoysall, C., Mahapatra, D., Sarada, R., Chauhan, V., Ramesh, A., 2012. Sustainability of large-scale algal biofuel production in India. *J Indian Inst Sci* 92, 63–98.

Huang, Y., Zheng, Y., Li, J., Liao, Q., Fu, Q., Xia, A., Fu, J., Sun, Y., 2018. Enhancing microalgae biofilm formation and growth by fabricating microgrooves onto the substrate surface. *Bioresour Technol* 261, 36–43.

Jain, S., 2020. Marine Diatoms. In: Jain, S. (Ed.), *Fundamentals of invertebrate palaeontology*. Springer, New Delhi, pp. 131–142.

Jiang, A., Ji, H., Liu, H., Zhu, H., Ai, G., Guo, X., 2020. Culture of benthic diatom Nitzschia sp. with macroalgae carriers and its application as feed of juveniles Stichopus japonicus. *Helgol. Mar. Res* 74, 11.

Jin, C., Yu, Z., Peng, S., Feng, K. E., Zhang, L., Zhou, X., 2018. The characterization and comparison of exopolysaccharides from two benthic diatoms with different biofilm formation abilities. *An. Acad. Bras. Cienc* 90, 1503–1519.

Jones, V. J. 2013. Diatom introduction. in: *Encyclopedia of Quaternary Science*, pp. 471–480.

Kardel, K. 2016. An analytical and experimental study on 3D-printed custom surfaces for benthic algal biofilms. in: Industrial and Systems Engineering, Vol. Doctor of Philosophy, Auburn University. Auburn, Alabana.

Karthikeyan, R., Krishnaraj, N., Selvam, A., Wong, J.W.-C., Lee, P. K. H., Leung, M. K. H., Berchmans, S., 2016. Effect of composites based nickel foam anode in microbial fuel cell using Acetobacter aceti and Gluconobacter roseus as a biocatalysts. *Bioresour Technol* 217, 113–120.

Kim, S., Moon, M., Kwak, M., Lee, B., Chang, Y. K., 2018. Statistical optimization of light intensity and CO_2 concentration for lipid production derived from attached cultivation of green microalga Ettlia sp. *Sci Rep* 8, 15390.

Krsmanovic, M., Biswas, D., Ali, H., Kumar, A., Ghosh, R., Dickerson, A. K., 2021. Hydrodynamics and surface properties influence biofilm proliferation. *Adv Colloid Interface Sci* 288, 102336.

Li, J., Stamato, M., Velliou, E., Jeffryes, C., Agathos, S. N., 2015. Design and characterization of a scalable airlift flat panel photobioreactor for microalgae cultivation. *J Appl Phycol* 27, 75–86.

Liang, Y., Kashdan, T., Sterner, C., Dombrowski, L., Petrick, I., Kröger, M., Höfer, R., 2015. Chapter 2 - Algal biorefineries. In: Pandey, A., Höfer, R., Taherzadeh, M., Nampoothiri, K. M., Larroche, C. (Eds.), *Industrial Biorefineries & White Biotechnology*. Elsevier, Amsterdam, pp. 35–90.

Litchman, E., 2007. Chapter 16- Resource competition and the ecological success of phytoplankton. In: Falkowski, P. G., Knoll, A. H. (Eds.), *Evolution of Primary Producers in the Sea*. Academic Press, Burlington, pp. 351–375.

Madigan, M. T., Martinko, J. M., Brock, T. D., 2006. *Brock biology of microorganisms*. Pearson Prentice Hall, Upper Saddle River, NJ.

Mantzorou, A., Ververidis, F., 2019. Microalgal biofilms: A further step over current microalgal cultivation techniques. *Sci Total Environ* 651, 3187–3201.

Marella, T. K., Datta, A., Patil, M. D., Dixit, S., Tiwari, A., 2019. Biodiesel production through algal cultivation in urban wastewater using algal floway. *Bioresour Technol* 280, 222–228.

Marella, T. K., López-Pacheco, I. Y., Parra-Saldívar, R., Dixit, S., Tiwari, A., 2020. Wealth from waste: Diatoms as tools for phycoremediation of wastewater and for obtaining value from the biomass. *Sci Total Environ* 724, 137960.

Marella, T. K., Tiwari, A., 2020. Marine diatom *Thalassiosira weissflogii* based biorefinery for co-production of eicosapentaenoic acid and fucoxanthin. *Bioresour Technol* 307, 123245.

Matos, Â.P., 2017. The impact of microalgae in food science and technology. *J. Am. Oil Chem. Soc* 94, 1333–1350.

Matsumoto, M., Nojima, D., Nonoyama, T., Ikeda, K., Maeda, Y., Yoshino, T., Tanaka, T., 2017. Outdoor cultivation of marine diatoms for year-round production of biofuels. *Mar Drugs* 15, 94.

Merlo, S., Gabarrell Durany, X., Pedroso Tonon, A., Rossi, S., 2021. Marine microalgae contribution to sustainable development. *Water* 13, 1373.

Mishra, M., Arukha, A., Bashir, T., Yadav, D., Prasad, G., 2017. All new faces of diatoms: Potential source of nanomaterials and beyond. *Front Microbiol* 8, 1–8.

Mishra, S., Mohanty, K., 2019. Comprehensive characterization of microalgal isolates and lipid-extracted biomass as zero-waste bioenergy feedstock: An integrated bioremediation and biorefinery approach. *Bioresour Technol* 273, 177–184.

Mizyed, A., 2021. Review on application of rotating biological contactor in removal of various pollutants from effluent. *Technium BioChemMed* 2, 41–61.

Narala, R., Garg, S., Sharma, K., Thomas-Hall, S., Deme, M., Li, Y., Schenk, P., 2016. Comparison of microalgae cultivation in photobioreactor, open raceway pond, and a two-stage hybrid system. *F Energy Res* 4, 29.

Nongmaithem, D., Tiwari, R., Goud, V., 2021. Cultivating Scenedesmus sp. on substrata coated with cyanobacterial-derived extracellular polymeric substances for enhanced biomass productivity: a novel harvesting approach. *Biomass Convers Biorefin* 13, 2971–2983. https://www.semanticscholar.org/paper/Cultivating-Scenedesmus-sp.-on-substrata-coated-for-Devi-Tiwari/b327d9b9e9987385d02c3458ce09c01a6717705f#citing-papers

Ozkan, A., Berberoglu, H., 2013. Adhesion of algal cells to surfaces. *Biofouling* 29, 469–482.

Pandey, L. K., Lavoie, I., Morin, S., Park, J., Lyu, J., Choi, S., Lee, H., Han, T., 2018. River water quality assessment based on a multi-descriptor approach including chemistry, diatom assemblage structure, and non-taxonomical diatom metrics. *Ecol Indic* 84, 140–151.

Patil, J. S., Anil, A. C., 2005. Biofilm diatom community structure: Influence of temporal and substratum variability. *Biofouling* 21, 189–206.

Pierre, G., Delattre, C., Dubessay, P., Jubeau, S., Vialleix, C., Cadoret, J. P., Probert, I., Michaud, P., 2019. What is in store for EPS microalgae in the next decade? *Molecules* 24, 1–25.

Podola, B., Li, T., Melkonian, M., 2017. Porous substrate bioreactors: A paradigm shift in microalgal biotechnology. *Trends Biotechnol* 35, 121–132.

Rosenhahn, A., Finlay, J. A., Pettit, M. E., Ward, A., Wirges, W., Gerhard, R., Callow, M. E., Grunze, M., Callow, J. A., 2009. Zeta potential of motile spores of the green alga Ulva linza and the influence of electrostatic interactions on spore settlement and adhesion strength. *Biointerphases* 4, 7–11.

Rosenhahn, A., Schilp, S., Kreuzer, H., Grunze, M., 2010. The role of "inert" surface chemistry in marine biofouling prevention. *Physical Chem Chem Phys* 12, 4275–4286.

Salta, M., Wharton, J., Blache, Y., Stokes, K., Briand, J.-F., 2013. Marine Biofilms on artificial surfaces: Structure and dynamics. *Environ Microbiol* 15, 2879–2893. https://ami-journals.onlinelibrary.wiley.com/doi/10.1111/1462-2920.12186

Sanz Smachetti, M., Sanchez Rizza, L., Coronel, C., Do Nascimento, M., Curatti, L., 2018. Microalgal biomass as an alternative source of sugars for the production of bioethanol. In: Kulia, A., Sharma, V. (Eds.), *Principles and Applications of Fermentation Technology*. Scrivener Publishing LLC, Beverly, United States, pp. 351–386.

Sato, R., Maeda, Y., Yoshino, T., Tanaka, T., Matsumoto, M., 2014. Seasonal variation of biomass and oil production of the oleaginous diatom Fistulifera sp. in outdoor vertical bubble column and raceway-type bioreactors. *J Biosci Bioeng* 117, 720–724.

Schnurr, P. J., Espie, G. S., Allen, D. G., 2013. Algae biofilm growth and the potential to stimulate lipid accumulation through nutrient starvation. *Bioresour Technol* 136, 337–344.

Sekar, R., Venugopalan, V. P., Satpathy, K. K., Nair, K. V. K., Rao, V. N. R. 2004. *Laboratory studies on adhesion of microalgae to hard substrates. Asian Pacific Phycology in the 21st Century: Prospects and Challenges*. Springer Netherlands, Dordrecht, pp. 109–116.

Sharp, C. E., Urschel, S., Dong, X., Brady, A. L., Slater, G. F., Strous, M., 2017. Robust, high-productivity phototrophic carbon capture at high pH and alkalinity using natural microbial communities. *Biotechnol Biofuels* 10, 84.

Shen, Y., Xu, X., Zhao, Y., Lin, X., 2014. Influence of algae species, substrata and culture conditions on attached microalgal culture. *Bioprocess Biosyst Eng* 37, 441–450.

Shen, Y., Zhang, H., Xu, X., Lin, X., 2015. Biofilm formation and lipid accumulation of attached culture of *Botryococcus braunii*. *Bioprocess Biosyst Eng* 38, 481–488.

Shniukova, E. I., Zolotareva, E. K., 2015. Diatom exopolysaccharides: A review. *Int J Algae* 17, 50–67.

Show, K.-Y., Lee, D.-J., 2014. Chapter 9 - Production of Biohydrogen from Microalgae. In: Pandey, A., Lee, D.-J., Chisti, Y., Soccol, C. R. (Eds.), *Biofuels from Algae*. Elsevier, Amsterdam, pp. 189–204.

Skjånes, K., Andersen, U., Heidorn, T., Borgvang, S. A., 2016. Design and construction of a photobioreactor for hydrogen production, including status in the field. *J Appl Phycol* 28, 2205–2223.

Song, F., 2016. Effects of substrate stiffness on bacterial biofilm formation. in: Chemical Engineering. *Vol. Doctor of Philosophy, Syracuse University*, pp. 453.

Song, F., Koo, H., Ren, D., 2015. Effects of material properties on bacterial adhesion and biofilm formation. *J Dent Res* 94, 1027–1034.

Soriano-Jerez, Y., López-Rosales, L., Cerón-García, M. C., Sánchez-Mirón, A., Gallardo-Rodríguez, J. J., García-Camacho, F., Molina-Grima, E., 2021. Long-term biofouling formation mediated by extracellular proteins in Nannochloropsis gaditana microalga cultures at different medium N/P ratios. *Biotechnol Bioeng* 118, 1152–1165.

Souza, M., Ferragut, C., 2012. Influence of substratum surface roughness on periphytic algal community structure in a shallow tropical reservoir. *Acta Limnol Braz* 24, 397–407.

Tang, J., Liu, B., Gao, L., Wang, W., Liu, T., Su, G., 2021. Impacts of surface wettability and roughness of styrene-acrylic resin films on adhesion behavior of microalgae Chlorella sp. *Colloids Surf B Biointerfaces* 199, 111522.

Telegdi, J., Shaban, A., Trif, L., 2017. 8 - Microbiologically influenced corrosion (MIC). In: El-Sherik, A. M. (Ed.), *Trends in Oil and Gas Corrosion Research and Technologies*. Woodhead Publishing, Boston, pp. 191–214.

Tong, C. Y., Chang, Y. S., Ooi, B. S., Chan, D. J. C., 2021. Physico-chemistry and adhesion kinetics of algal biofilm on polyethersulfone (PES) membrane with different surface wettability. *J Environ Chem Eng* 9, 106531.

Tong, C. Y., Derek, C. J. C., 2021a. Biofilm formation of benthic diatoms on commercial polyvinylidene fluoride membrane. *Algal Res* 55, 102260.

Tong, C. Y., Derek, C. J. C., 2022. Membrane surface roughness promotes rapid initial cell adhesion and long term microalgal biofilm stability. *Environ Res* 206, 112602.

Tong, C. Y., Derek, C. J. C., 2021b. The role of substrates towards marine diatom *Cylindrotheca fusiformis* adhesion and biofilm development. *J Appl Phycol* 33, 2845–2862.

Tong, C. Y., Lew, J. K., Derek, C. J. C., 2022. Algal extracellular organic matter pre-treatment enhances microalgal biofilm adhesion onto microporous substrate. *Chemosphere* 307, 135740.

Underwood, G., Boulcott, M., Raines, C., Waldron, K., 2004. Environmental effects on exopolymer production by marine benthic diatoms: Dynamics, changes in composition, and pathways of production. *J Phycol* 40, 293–304.

Wang, J. K., Seibert, M., 2017. Prospects for commercial production of diatoms. *Biotechnol Biofuels* 10, 1–13.

Wood, J. L., Takemoto, J. Y., Sims, R. C., 2022. Rotating algae biofilm reactor for management and valorization of produced wastewater. *F. Energy Res* 10. https://www.frontiersin.org/articles/10.3389/fenrg.2022.774760/full

Xu, L., Weathers, P. J., Xiong, X.-R., Liu, C.-Z., 2009. Microalgal bioreactors: Challenges and opportunities. *Engn Life Sci* 9, 178–189.

Zhang, J., Huang, J., Say, C., Dorit, R. L., Queeney, K. T., 2018a. Deconvoluting the effects of surface chemistry and nanoscale topography: *Pseudomonas aeruginosa* biofilm nucleation on Si-based substrates. *J. Colloid Interface Sci* 519, 203–213.

Zhang, Q., Yu, Z., Jin, S., Liu, C., Li, Y., Guo, D., Hu, M., Ruan, R., Liu, Y., 2020. Role of surface roughness in the algal short-term cell adhesion and long-term biofilm cultivation under dynamic flow condition. *Algal Res* 46, 101787.

Zhang, X., Yuan, H., Jiang, Z., Lin, D., Zhang, X., 2018b. Impact of surface tension of wastewater on biofilm formation of microalgae Chlorella sp. *Bioresour Technol* 266, 498–506.

Zhao, Z., Muylaert, K., Szymczyk, A., Vankelecom, I. F. J., 2021. Enhanced microalgal biofilm formation and facilitated microalgae harvesting using a novel pH-responsive, crosslinked patterned and vibrating membrane. *Chem Eng J* 410, 127390.

Zhou, H., Sheng, Y., Zhao, X., Gross, M., Wen, Z., 2018. Treatment of acidic sulfate-containing wastewater using revolving algae biofilm reactors: Sulfur removal performance and microbial community characterization. *Bioresour Technol* 264, 24–34.

Zhuang, L.-L., Azimi, Y., Yu, D., Wang, W.-L., Wu, Y.-H., Dao, G.-H., Hu, H.-Y., 2016. Enhanced attached growth of microalgae Scenedesmus. LX1 through ambient bacterial pre-coating of cotton fiber carriers. *Bioresour Technol* 218, 643–649.

Index